Advanced Phase Change Materials for Thermal Storage

Advanced Phase Change Materials for Thermal Storage

Editor

Rocío Bayón

MDPI • Basel • Beijing • Wuhan • Barcelona • Belgrade • Manchester • Tokyo • Cluj • Tianjin

Editor
Rocío Bayón
Thermal Storage and Solar Fuels
Unit, CIEMAT-PSA
Spain

Editorial Office
MDPI
St. Alban-Anlage 66
4052 Basel, Switzerland

This is a reprint of articles from the Special Issue published online in the open access journal *Applied Sciences* (ISSN 2076-3417) (available at: https://www.mdpi.com/journal/applsci/special_issues/phase_change_material_thermal_storage).

For citation purposes, cite each article independently as indicated on the article page online and as indicated below:

LastName, A.A.; LastName, B.B.; LastName, C.C. Article Title. *Journal Name* **Year**, *Volume Number*, Page Range.

ISBN 978-3-0365-0864-1 (Hbk)
ISBN 978-3-0365-0865-8 (PDF)

Cover image courtesy of Rocío Bayón.

Contents

About the Editor

Rocío Bayón received an M.Sc. degree in Chemistry at the University of Valladolid and a Ph.D. degree at the University Autónoma of Madrid. She was a postdoctoral fellow at the Hahn-Meitner-Institut of Berlin (Germany) and worked as research engineer at IMRA Europe in Sophia Antipolis (France). At present, she works as senior scientist in the Thermal Energy Storage Unit of CIEMAT-PSA in Madrid. She has been working in different scientific areas related to materials and systems for renewable energy applications such as thin film solar cells, selective absorbers and thermal energy storage. She has participated in many national and international research projects, international conferences and is the co-author of about 40 scientific papers. She collaborates as a teacher in various university masters programs and participates in technological collaboration platforms of the International Energy Agency related to thermal storage.

Editorial

Special Issue "Advanced Phase Change Materials for Thermal Storage"

Rocío Bayón

Thermal Energy Storage Unit, CIEMAT-PSA, Av. Complutense 40, 28040 Madrid, Spain; rocio.bayon@ciemat.es; Tel.: +34-913466048

Citation: Bayón, R. Special Issue "Advanced Phase Change Materials for Thermal Storage". *Appl. Sci.* **2021**, *11*, 1390. https://doi.org/10.3390/app11041390

Received: 25 January 2021
Accepted: 30 January 2021
Published: 4 February 2021

Publisher's Note: MDPI stays neutral with regard to jurisdictional claims in published maps and institutional affiliations.

1. Introduction

Thermal energy storage using phase change materials (PCMs) is a research topic that has attracted much attention in recent decades. This is mainly because the potential use of PCMs as latent storage media not only covers renewable energy and building efficiency applications, but also the temperature control of electronic devices, batteries and even clothes. Although a number of companies worldwide are producing a variety of PCMs, advanced materials with improved properties and new latent storage concepts are required to better meet the specific requirements of different applications. Moreover, the development of common validation procedures for PCMs is an important issue that should be addressed in order to achieve commercial deployment and implementation of these kinds of materials in latent storage systems.

2. Advanced Phase Change Materials for Thermal Storage

The key subjects included in this special issue were related not only to materials in terms of new PCM formulations and concepts, validation and assessment procedures, characterization and simulation, but also to PCM applications in terms of implementation and testing in storage prototypes, innovative approaches and the simulation of novel storage modules for latent heat. Despite COVID-19 crises and lockdowns in most countries, there were still six papers submitted to this special issue, and five of them were accepted, which proves the quality of the research and the strong interest in the field of latent heat storage. In the following paragraphs, a summary of these papers with their most relevant contributions is presented.

The first paper included in this issue dealt with a procedure for selecting the appropriate PCM for two kinds of innovative compact energy storage systems implemented in residential buildings: the Mediterranean (MED) concept, intended for space cooling, and the continental (CON) concept, used for heating and domestic water [1]. The selection methodology consists of a qualitative decision matrix, which uses some common features of PCMs to assign an overall score to each material so that different options can be compared. The most important PCM features to be considered in the decision matrix for material selection are the melting enthalpy and temperature range, availability, cost and, in the case of the CON concept, the maximum working temperature range. Apart from the qualitative results, the authors made an experimental characterization of the best candidates, consisting of differential scanning calorimetry (DSC) and thermogravimetric analysis (TGA), before making a final decision. This selection process led to various possible candidates, but two commercial PCMs were selected as the most promising ones: RT4 (T_{melt} = 4 °C) for the MED cooling system and RT64HC (T_{melt} = 64 °C) for the CON system providing house heating and domestic hot water.

The main contribution of this paper is that it provides a simple, quick tool for prescreening a PCM before being implemented in any application and selecting at least the most promising candidates to be included in further validation tests.

The second paper presented a methodology that allowed comparing latent heat energy storage (LHES) modules with different designs with respect to their compactness and heat transfer performance [2]. Nowadays, many novel and promising heat exchanger designs and concepts have emerged, aiming to enhance the heat transfer inside LHES devices. However, the wide range of experimental conditions that can be found in the literature for their characterization makes it difficult to compare their performance. In light of this, the methodology described in the paper established just two key performance indicators (KPIs) which were minimally influenced by the experimental conditions: the compactness degree (ΦPCM) and a normalized heat transfer performance coefficient (NHTPC). In the paper, these KPIs were calculated for several LHES units already reported in the literature, allowing a leveled performance comparison with regard to operating conditions at different scales, while characteristics like geometry, structural materials and PCMs remained intrinsic. The robustness of the proposed KPIs was confirmed for units at different size scales by varying the heat transfer fluid mass flow rates and temperature levels. The evaluation procedure was applied to various LHES systems, and the most promising designs for different applications were identified and discussed. The authors clarified that the storage units analyzed were application-oriented and, in most cases, a high heat transfer rate was not a requirement, which led to low values for the KPIs. However optimized versions of the evaluated LHES systems are expected to deliver considerably higher performance indicators. Hence, the conclusions of this analysis should be considered as preliminary pictures of the general potential associated to each technological approach The main contribution of this paper is that the methodology proposed is expected to open new paths in LHES research by allowing the leveled-ground comparison of technologies among different studies, facilitating the evaluation and selection of the most suitable design or designs for a specific storage application.

The third paper accepted in this special issue introduced the use of PCMs for the thermal management of lithium-ion batteries (LIBs), since temperature is an important factor affecting the working efficiency and service life of these devices [3]. In this work the authors studied the thermal performance of two commercial batteries (Sony and Sanyo) under different working conditions: extreme conditions (inside a closed aluminum tube), natural convection cooling and PCM cooling with and without heat dissipation fins. The PCM used was a composite of wax with expanded graphite (T_{melt} = 52 °C) and the experimental results showed that the PCM was able to absorb some of the heat produced during both the charge and discharge processes and, hence, effectively reduce the temperature and keep the battery capacity stable. In fact, the tests performed at different discharge rates showed that the temperature decrease attained under PCM cooling was much higher for the Sanyo LIB (between 4.7 °C and 12.8 °C) than for the Sony LIB (between 1.1 °C and 2 °C), in both cases being compared with the natural convection experiments The temperature reduction impact on the Sanyo LIB was stronger because this battery generated more heat due to its larger storage capacity. As for future developments, the authors suggested that further optimization of LIB thermal management in terms of surface temperature reduction could be achieved if a PCM with a higher latent heat was combined with heat dissipation fins. In my opinion, the most interesting contribution of this paper is that electrical storage and thermal storage working together can improve the performance efficiency of energy storage systems, which is one of the main challenges to be addressed and solved in future energetic scenarios.

The fourth paper resulted from the collaboration of several institutions and was developed within the framework of Annex 33/SHC Task 58 Material and Component Development for Compact Thermal Energy Storage, a joint working group of the Energy Storage (ES) and Solar Heating and Cooling (SHC) Technology Collaboration Programmes of the International Energy Agency (IEA) [4]. It consisted of a survey with a detailed description of the experimental devices present in those institutions and used for investigating the long-term stability and performance of PCMs under application conditions [5] In fact, an important prerequisite to select a reliable material for thermal energy storage applications is to investigate its performance under real working conditions. In the case of solid–liquid PCMs, the long-term performance in terms of the melting and solidification processes should be ensured along the lifetime of the storage system, taking into account the conditions of the intended application. In this work, the different institutions presented up to 18 experimental set-ups and devices that allowed for performing thermal tests (cycling and constant temperature) not only for conventional PCMs, but also for the ones with stable supercooling, as well as phase change slurries (PCSs). Moreover, the paper introduced appropriate methods to investigate possible degradation mechanisms of PCMs. Considering the diversity of the devices and the wide range of experimental parameters, further work toward a standardization of PCM stability testing is strongly recommended. The main contribution of this paper is that it puts together many experimental facilities currently in use and the know-how of the corresponding institutions for assessing the long-term performance of PCMs, which certainly is a key issue for the commercial implementation of LHTS systems.

The last paper of this special issue presented a compact model of latent heat thermal storage (LHTS) for its integration in multi-energy systems [6]. In this way, the study developed a new modeling approach to quickly characterize the dynamic behavior of an LHTS unit. The thermal power released or absorbed by an LHTS module was expressed only as a function of the current and the initial state of charge. Moreover, the model also allowed for simulating even the partial charge and discharge processes. In general, the results were quite accurate when compared with a 2D finite volume model, with the advantage of the computational effort being much lower. Due to its simplicity, this model can be used to investigate optimal LHTS control strategies at the system level. In light of this, the authors implemented it in two relevant case studies: the reduction of the morning thermal power peak in district heating systems and the optimal energy supply schedule in multi-energy systems. However, this study describes the functioning of the LHTS unit at the system level only on the basis of numerical results. Hence, future work should also test the LHTS unit in a real case application to better quantify the model uncertainties.

The main contribution of this paper is the development of a simple model for LHTS modules that can be implemented in the simulation of multi-energy systems, although the model uncertainties still remain unquantified since it should be validated with data obtained from real applications.

Funding: This research received no external funding.

Institutional Review Board Statement: Not applicable.

Informed Consent Statement: Not applicable.

Data Availability Statement: Not applicable.

Conflicts of Interest: The author declares no conflict of interest.

eferences

Zsembinszki, G.; Fernández, A.G.; Cabeza, L.F. Selection of the Appropriate Phase Change Material for Two Innovative Compact Energy Storage Systems in Residential Buildings. *Appl. Sci.* **2020**, *10*, 2116. [CrossRef]
Delgado-Diaz, W.; Stamatiou, A.; Maranda, S.; Waser, R.; Worlitschek, J. Comparison of Heat Transfer Enhancement Techniques in Latent Heat Storage. *Appl. Sci.* **2020**, *10*, 5519. [CrossRef]
Chen, M.; Zhang, S.; Wang, G.; Weng, J.; Ouyang, D.; Wu, X.; Zhao, L.; Wang, J. Experimental Analysis on the Thermal Management of Lithium-Ion Batteries Based on Phase Change Materials. *Appl. Sci.* **2020**, *10*, 7354. [CrossRef]
IEA ES Annex 33/SHC Task 58 "Material and Component Development for Compact Thermal Energy Storage". Available online: https://task58.iea-shc.org/ (accessed on 25 January 2021).
Rathgeber, C.; Hiebler, S.; Bayón, R.; Cabeza, L.F.; Zsembinszki, G.; Englmair, G.; Dannemand, M.; Diarce, G.; Fellmann, O.; Ravotti, R.; et al. Experimental Devices to Investigate the Long-Term Stability of Phase Change Materials under Application Conditions. *Appl. Sci.* **2020**, *10*, 7968. [CrossRef]
Colangelo, A.; Guelpa, E.; Lanzini, A.; Mancò, G.; Verda, V. Compact Model of Latent Heat Thermal Storage for Its Integration in Multi-Energy Systems. *Appl. Sci.* **2020**, *10*, 8970. [CrossRef]

Article

Compact Model of Latent Heat Thermal Storage for Its Integration in Multi-Energy Systems

Alessandro Colangelo *, Elisa Guelpa, Andrea Lanzini, Giulia Mancò and Vittorio Verda

Department of Energy, Politecnico di Torino, 10129 Torino, Italy; elisa.guelpa@polito.it (E.G.); andrea.lanzini@polito.it (A.L.); giulia.manco@polito.it (G.M.); vittorio.verda@polito.it (V.V.)
* Correspondence: alessandro.colangelo@polito.it

Received: 30 October 2020; Accepted: 14 December 2020; Published: 16 December 2020

Featured Application: Dynamic simulation of Latent Heat Thermal Storage at system level; optimization of control strategies in multi-energy systems; investigation of Demand Side Management strategies.

Abstract: Nowadays, flexibility through energy storage constitutes a key feature for the optimal management of energy systems. Concerning thermal energy, Latent Heat Thermal Storage (LHTS) units are characterized by a significantly higher energy density with respect to sensible storage systems. For this reason, they represent an interesting solution where limited space is available. Nevertheless, their market development is limited by engineering issues and, most importantly, by scarce knowledge about LHTS integration in existing energy systems. This study presents a new modeling approach to quickly characterize the dynamic behavior of an LHTS unit. The thermal power released or absorbed by a LHTS module is expressed only as a function of the current and the initial state of charge. The proposed model allows simulating even partial charge and discharge processes. Results are fairly accurate when compared to a 2D finite volume model, although the computational effort is considerably lower. Summarizing, the proposed model could be used to investigate optimal LHTS control strategies at the system level. In this paper, two relevant case studies are presented: (a) the reduction of the morning thermal power peak in District Heating systems; and (b) the optimal energy supply schedule in multi-energy systems.

Keywords: latent heat thermal storage; pcm; 0D dynamic model; multi-energy system; district heating; thermal network

---

1. Introduction

The thermal energy storage is a key element for increasing the operational flexibility of energy systems, especially when these are supplied by renewable energy sources. When the space available for storage technologies is limited, such as in urban contexts, Latent Heat Thermal Storage (LHTS) through the melting/solidification of Phase Change Materials (PCMs) can reduce the occupied volume up to five times with respect to sensible thermal storages to reach the same energy content. Therefore, an increasing amount of attention is focused on LHTS systems in order to advance their Technology Readiness Level (TRL) and market development. Currently, LHTS technologies have scarcely been introduced to market, and their TRL is lower than seven [1], which indicates that this technology has generally been demonstrated only in operational environments.

In this framework, numerous LHTS modeling approaches have been proposed in the existing literature. LHTSs are commonly analyzed through detailed numerical models, which, in few cases, are validated by experimental results. In fact, the intrinsic physical complexity of the phase change phenomenon can be easily simulated by numerical models [2]. Most of the models are focused on LHTSs design optimization with the aim of enhancing the heat transfer between the PCM and the heat transfer fluid (HTF). As reported in [3], several enhancement techniques can be pursued. Among these, numerous works investigated the optimal sizing of HTF tube fins. For instance, Niyas et al. [4] performed a numerical study of a shell-and-tube LHTS prototype filled with PCM salts with a full 3D model. Using commercial software based on a finite element scheme, they determined the optimal number of HTF tubes and longitudinal fins per tube that minimizes the overall discharging time. They also monitored a few performance parameters of the identified configuration, such as the melting fraction, charging/discharging times, outlet HTF temperature, and energy stored. Then, results were validated with experiments [5] showing good agreement. Sciacovelli et al. [6] adopted a shape optimization strategy for Y-shaped fins with one or two bifurcations in order to enhance the performance of a shell-and-tube LHTS. A two-dimensional cross-section model, solving the underlying equations with a commercial finite volume code, was implemented. The fins optimization was performed requiring the complete discharge of the LHTS unit both in a short and long-time interval. A further application of CFD modeling to LHTS fins design optimization was presented by Pizzolato et al. [7]. They extended the investigation of the fins design through the combination of topology optimization techniques and ad hoc finite element two-dimensional and three-dimensional models. Here, only heat conduction is considered. Then, the same authors increased their model complexity by including also natural convection [8].

Esapour et al. [9] developed a detailed 3D in-house code to study the performance of an annular LHTS with multiple tubes. More specifically, they determined the effect of the arrangement and number of tubes on the liquid fraction and melting time. Then, in another work [10], the same model was also adopted to investigate the influence of variable operational parameters such as HTF inlet temperature and flow rate on the LHTS performance. Similarly, Saddegh et al. [11] compared the thermal behavior of a shell-and-tube LHTS using a 2D conduction–convection heat transfer model for two different LHTS orientations (vertical and horizontal).

In an attempt to simplify the simulation of LHTS systems, Neumann et al. [12] coupled a 1D model for the heat exchanger tubes and a reduced 3D model for the PCM and fins. This model can be used to improve the geometric and operational parameters of fin-and-tubes LHTSs with a limited computational effort, although the PCM parameters should be carefully calibrated. A different simplification approach was followed by Parry et al. [13]: a 3D computational model for a shell-and-tube LHTS was calibrated with experimental results. This was successively reduced to a one-dimensional radial model by employing an effective diffusivity technique. According to the authors, this model is suited to predict performance over long-time spans. An original simple model was proposed by Tay et al. [14]. This is based on the effectiveness-number of transfer units, and it was calibrated on a tube-in-tank salt-based PCM system with radial circular fins. The correlation proposed by the authors is semi-empirical, and it can be used for sizing and optimizing LHTS systems. A further interesting simplification methodology was suggested by Johnson et al. [15]. First, they analyzed the heat transfer properties of a tube assembly cross-section through a 2D finite volume model. Then, they produced a 1D radial model in Dymola for the ensemble of PCM and fins able to yield comparable results. Finally, they coupled this latter model with a 1D axial model for the HTF again in Dymola and studied the time evolution of the thermal heat flux for different configurations.

Nevertheless, although a few studies investigated the integration of LHTS in existing heating systems, little attention has been devoted to the analysis of LHTSs operating conditions and partial charge/discharge. For instance, Colella et al. [16] analyzed the behavior of an LHTS unit in a District Heating (DH) substation through a 2D finite volume model. Xu et al. [17] studied the performance of an encapsulated LHTS unit in a residential heating system in Sweden by means of a finite element model. Johnson et al. [15] designed an LHTS for the production of high-temperature steam in a cogeneration plant using the simplified model previously mentioned. However, in order to facilitate LHTS market development, their optimal integration in energy systems is essential. Dynamic models that are fast and accurate could positively support the study of LHTSs at a system level.

In this work, a compact 0D model for a modular shell-and-tube LHTS is proposed. The model is able to predict the thermal power released or absorbed by the assembly as a function of its state of charge. Furthermore, it can represent partial charge and discharge operations. Therefore, this model can be easily deployed for simulations at a system level in order to investigate optimal control strategies or demand side management (DSM). Indeed, the proposed model is here applied to two relevant case studies regarding the integration of LHTS in DH and multi-energy systems.

2. Materials and Methods

In order to study the optimal integration of different energy technologies at the building level, a reliable representation of each energy asset is essential. In this framework, an LHTS model for the simulation of the operating conditions provides quick results while retaining an acceptable physical description. Considering the physical complexity of the phase change in PCMs and the heat transfer in shell-and-tube configurations, a two-step model simplification was adopted. First, a 2D numerical model was developed in Ansys Fluent (2020 R1). Then, a 0D mathematical description was adopted, fitting the results obtained from the former numerical model with an exponential function. The LHTS system used as a reference is a vertical shell-and-tube type, where finned tubes are crossed by water as heat transfer fluid (HTF) and its shell encloses the PCM.

2.1. D Detailed Model

Figure 1 shows the computational domain of the initial detailed model. Due to the simultaneous presence of a solid and liquid phase in the PCM, both heat conduction and natural convection are relevant heat transfer phenomena in LHTS systems. Thus, a full 3D numerical domain is generally recommended. However, the characteristics of the configuration analyzed in this study allow a substantial domain reduction. In fact, as claimed by Vogel et al. [18], compact LHTS designs hinder the development of natural convection phenomena, which are the main aspects responsible for problem asymmetries. According to [18], the Rayleigh number (Ra_W) and LHTS aspect ratio (A) can be used to determine the relevance of the wall heat flux due to natural convection in the liquid PCM. This correlation asserts that the flow is dominated by natural convection only if $\frac{Ra_W}{A} \geq 500$. Regarding the LHTS configuration proposed in this analysis, the above-mentioned correlation yields $\frac{Ra_W}{A} = 374$. Therefore, the heat flux due to natural convection is likely to be negligible both during charging and discharging phases if compared to conduction. Hence, the computational domain is reduced to the horizontal cross-section plane. Here, the temperature gradient is expected to be larger compared to the axial direction [19], and heat conduction is dominant. The numerical domain is also further reduced, taking advantage of the symmetry created by the fins. Then, the physical problem is modeled using the finite volume method implemented in the commercial code Ansys Fluent.

Regarding the materials, the fins and tubes are made of aluminum, while the PCM is assumed to be RT70HC manufactured by Rubitherm GmbH [20]. A summary of the PCM properties is reported in Table 1. The enthalpy–porosity technique is used for modeling the melting and solidification process [21]. Combining this method with the hypothesis of negligible natural convection, the only governing equation to be solved is the energy equation, as expressed in Equation (1).

$$\frac{\partial}{\partial t}(\rho H) - \nabla\cdot(k\nabla T) = 0 \tag{1}$$

where H is the sum of the sensible enthalpy and latent heat of the material.

Adiabatic boundary conditions are applied on the external face (Γ_3 in Figure 1), while symmetry is considered on Γ_2. The heat transfer with the water in the pipes is modeled with an implicit Robin boundary condition on Γ_1, as shown by Equations (2)–(4). T_{ref} represents the average temperature between the water inlet (T_{in}) and outlet (T_{out}) temperatures. However, T_{out} depends on the efficacy of the heat transfer between the flowing HTF and the rest of the LHTS (i.e., the ensemble of fins and PCM). Therefore, if T_{out} is expressed as a function of the heat flux at the wall ($\dot{q}_{wall}$), Fluent solves Equation (2) iteratively. The heat transfer coefficient h_{conv} is computed using the Dittus–Boelter correlation (5) and the Nusselt number (6). In (5), n = 0.4 for discharge or n = 0.3 for charge. This correlation was preferred over the implicit one proposed by Sieder and Tate due to its simplicity and acceptable uncertainty. Substantially, the simulated domain represents an average cross-section of the whole LHTS unit.

$$\dot{q}_{wall} = h_{conv}\left(T - T_{ref}\right) \tag{2}$$

$$T_{ref} = \frac{(T_{in} + T_{out})}{2} \tag{3}$$

$$T_{out} = T_{in} + \frac{L_{tube}\pi d_i}{Gc_{p_w}}\dot{q}_{wall} \tag{4}$$

$$Nu = 0.023Re^{\frac{4}{5}}Pr^{n} \tag{5}$$

$$h_{conv} = \frac{Nu * k_w}{d_i} \tag{6}$$

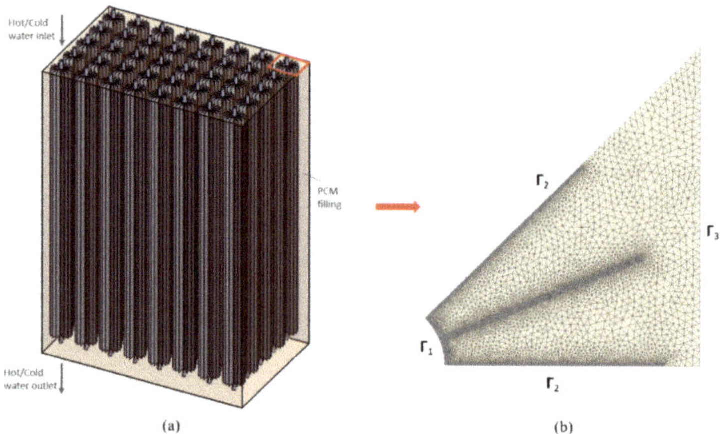

Figure 1. (**a**) Three-dimensional (3D) Latent Heat Thermal Storage (LHTS) shell-and-tube concept; (**b**) 2D computational domain.

Table 1. Phase Change Materials (PCM) physical properties.

PCM Physical Property	Value
Density (ρ) [kg/m^3]	880 (constant)
Specific Heat [J/(kg*K)]	2000
Thermal Conductivity (k) [W/(m*K)]	0.2
Latent Heat [kJ/kg]	214
Solidus Temperature [°C]	69
Liquidus Temperature [°C]	71

The numerical model follows the same approach adopted by Sciacovelli et al. [6], who validated their 2D cross-section model for PCM solidification against experimental data. For further details, the interested reader is referred to [6]. The computational grid consists of a non-structured mesh made of 8278 cells with an average cell size of 3.5×10^{-4} m. The selected mesh proved to be sufficiently fine not to influence the results. A Second Order Upwind scheme is used for the spatial discretization of the energy equation together with the Least Squares Cell Based method for gradient calculation.

The convergence is reached when residuals are lower than 10^{-8} for the energy equation. On the other hand, the transient nature of the problem is approached with a Second Order Implicit Euler method, with a time-step of 1 s. The selected value proved to be sufficiently fine not to influence the results.

2.2. D Compact Model

Considering the LHTSs dynamics, the thermal power released or absorbed by LHTSs varies over time, as the thermal conductivity of the liquid and solid material is different. Indeed, as the solidification or liquefaction of the PCM proceeds, the power exchanged decreases, since the melting front is further from the fins. For this reason, it was assumed that the most relevant parameter influencing the thermal power input/output of an LHTS is represented by its state of charge (*SOC*). The *SOC* can be defined as the ratio between the actual energy stored inside the LHTS (E_{stored}) and its reference value when fully charged (7).

$$SOC = \frac{E_{stored}(t) - E_{min}}{E_{max} - E_{min}} \tag{7}$$

where E_{max} is the energy contained in the PCM when the LHTS is fully charged. Similarly, E_{min} is the energy contained in the PCM when fully discharged.

A mathematical correlation between the current state of charge of the LHTS (*SOC*) and its thermal power input/output was identified using the results of the 2D model as shown by Equations (8)–(10). These were obtained observing the shape of the 2D model results and searching for their best fit through the sum of exponential functions. Each expression refers to a different operational phase (i.e., charge, discharge, or idle phase).

$$\dot{Q}_{dis} = Ae^{B*SOC_n} + Ce^{D*SOC_n} + K(1 - SOC_0)e^{-(\frac{SOC_n - E}{F})^2} \tag{8}$$

$$\dot{Q}_{chrg} = Ae^{B*SOC_n} + Ce^{D*SOC_n} + K * SOC_0 e^{-(\frac{SOC_n - E}{F})^2} \tag{9}$$

$$\dot{Q}_{idle} = 0 \tag{10}$$

SOC_n is the normalized state of charge, and it is equal to $\left(\frac{SOC}{SOC_0}\right)$ during discharge and $\left(\frac{SOC - SOC_0}{1 - SOC_0}\right)$ during charge, SOC_0 is the initial state of charge, $\dot{Q}_{dis}$ is the thermal power released by each tube during discharge, $\dot{Q}_{chrg}$ is the thermal power requested by each tube during the charging phase, and $\dot{Q}_{idle}$ is the thermal power when the LHTS is not operated. Therefore, depending on the operational phase of the LHTS, the value of the thermal power released or requested by each LHTS tube can be expressed as follows (11):

$$\dot{Q}_{LHTS} = x_1\dot{Q}_{dis} - x_2\dot{Q}_{dis} + x_3\dot{Q}_{idle} \tag{11}$$

where $x_1, x_2, x_3 \in [0; 1]$ and $x_1 + x_2 + x_3 = 1$.

Equations (8) and (9) were obtained fitting with exponential and Gaussian functions the LHTS thermal power and states of charge resulting from the simulations of the 2D model. The initial conditions of these simulations were, respectively, a fully charged LHTS and a fully discharged LHTS. Moreover, assuming that the thermal losses from the PCM to the HTF are negligible, the thermal power is always zero during the idle phase. Although these expressions were retrieved considering a specific LHTS design with fixed operating parameters (such as water mass flow rate and temperatures), the proposed methodology can be easily tailored to different LHTS concepts and operating conditions. Indeed, the evolution of the LHTS heat rate highlighted by several studies [12,15,22] is comparable with the one expressed by Equations (8) and (9).

3. Case Study

3.1. LHTS System Description

The LHTS system considered in this study is a vertical shell-and-tube type, with a height of 1.5 m. Each tube has an inner diameter of 19.05 mm (6/8″) and is equipped with 16 longitudinal fins whose extension is 30 mm (thickness: 1 mm). Instead, the interaxial distance between each tube is assumed to be 91 mm. The PCM is characterized by an average phase change temperature of 70 °C, as reported in Table 1. The HTF inlet temperature is assumed to be 75 °C during the charging phase and 48 °C during the discharge. These values were selected analyzing the working temperatures of the heating system of the building described in Section 3.2. The HTF mass flow rate is 0.168 kg/s per tube. This value was obtained fixing an average HTF velocity of 0.6 m/s. Finally, the size of the LHTS system (in terms of stored energy) is not fixed a priori. As a matter of fact, the LHTS can be conceived as a modular system composed by the aggregation of several tubes (n_{tubes}) and their associated PCM depending on the considered application. Each unit (tube + PCM) is able to store 2637.2 kJ, considering both the sensible and latent heat content of the PCM (at the selected temperatures).

3.2. Model Application to Thermal Energy Networks

Among the strategies to improve the operational management of DH networks, the adoption of distributed thermal storage systems at the building level is gaining interest as a measure to shave peak demands [23]. Therefore, the first case study considered in this article is the reduction of the peak thermal request from a substation of the DH network of Torino (Italy), which is characterized by a climate zone E. This substation belongs to a building known as the Energy Center, which is a research center owned by Politecnico di Torino. Thanks to its monitoring system, the building energy data are available with a frequency of 15 min. For this case study, a representative day of January 2020 is taken as a reference. As can be seen in Figure 2, the building heat demand has a common shape characterized by a morning peak (from 6:30 to 9 am) and a rather steady thermal request during the remainder of the day. During the peak hours, the amount of heat requested to the DH network is 3100 MJ, which represents the 27.1% of the overall heat demand within the same day.

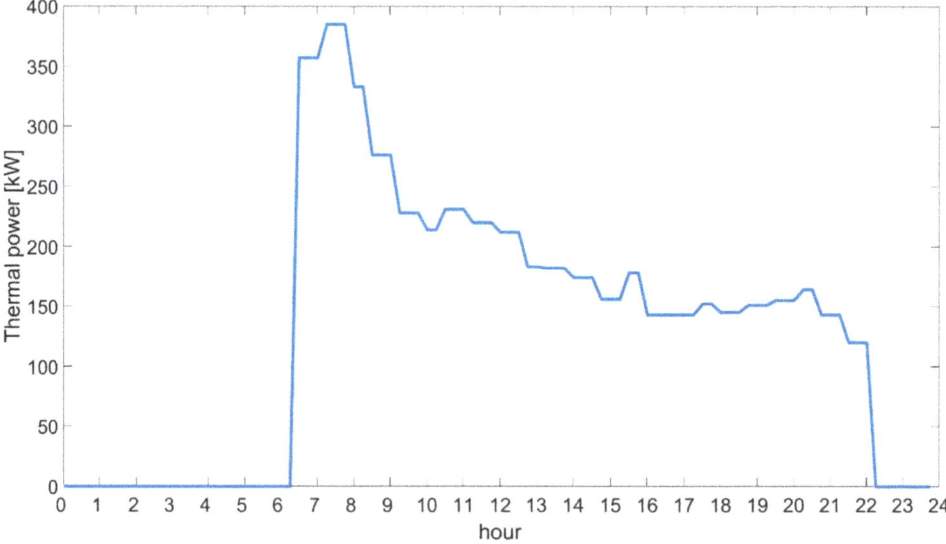

Figure 2. Building heat demand (15-min average thermal power).

Overall, an LHTS system composed of 300 units (791 MJ) is needed in order to achieve an energy demand reduction of 25.5% during the peak. However, due to the high power ceased or absorbed by each LHTS unit at the beginning of its operation, the overall storage system is subdivided into six bundles made of 50 units (tube + PCM) as depicted in Figure 3. These bundles are sequentially activated with a delay of few minutes both during charge (in the night) and discharge (in the morning peak). Moreover, each bundle follows the subsequent duty cycle: it is charged up to $SOC = 0.97$ and discharged down to $SOC = 0.02$. Further partial charge and discharge cycles are not considered for this case study. This is done since the building heat demand is rather constant for the rest of the day.

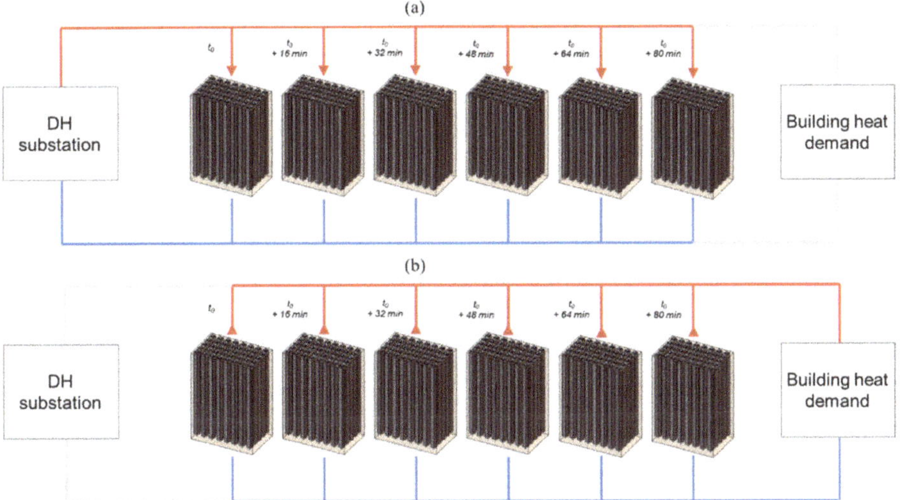

Figure 3. LHTS charge (**a**) and discharge (**b**) schedule for District Heating (DH) case study.

3.3. Model Application to Multi-Energy Systems

This section shows the application of the 0D model for the achievement of the best operation in a multi-energy system context, as schematized in Figure 4. The multi-energy system is conceived to supply heat, electricity, and cold to a block of flats with 15 dwellings. The production system is characterized by a group of several technologies. In this context, the adoption of a thermal storage provides significant benefits to improve the system flexibility. In particular, an LHTS can be extremely advantageous in buildings, since it requires small installation volumes. In order to easily model the time-dependent behavior of the LHTS, the 0D model is adopted.

The most convenient technologies to be operated are selected by a Mixed Integer Non-Linear Programming optimization tool that includes the 0D LHTS model. The optimization algorithm allows achieving the best operation for the multi-energy system. Therefore, its output is the optimal time evolution of (a) the power from the production/conversion technologies; (b) the power purchase from the grid; and (c) the released/absorbed power in the storages.

The optimization model includes constraints such as (a) the overall produced and purchased energy vectors must be equal to the loads and (b) the energy production of each technology must not exceed its maximum capacity (the same for the conversion technologies and the storages). Moreover, additional constraints are set to model the LHTS. The maximum thermal power that can be absorbed/released by the LHTS is imposed at each time step in relation to the outcome of the LHTS 0D model.

Concerning the energy production/conversion, both traditional and innovative technologies are considered to be installed. Among the traditional technologies that can be adopted are a heat-only boiler (HOB) for heat production, an electric chiller (EHP) for cold production, and the grid connection for the electricity/heat supply. Among the innovative technologies are a micro-cogeneration system (for combined heat and power production) fed by natural gas, an absorption chiller, fuel cells, photovoltaics (PV), thermal solar, and wind turbine. Furthermore, a series of storages (hot, cold, and electric) provide a significant level of flexibility for the multi-energy system.

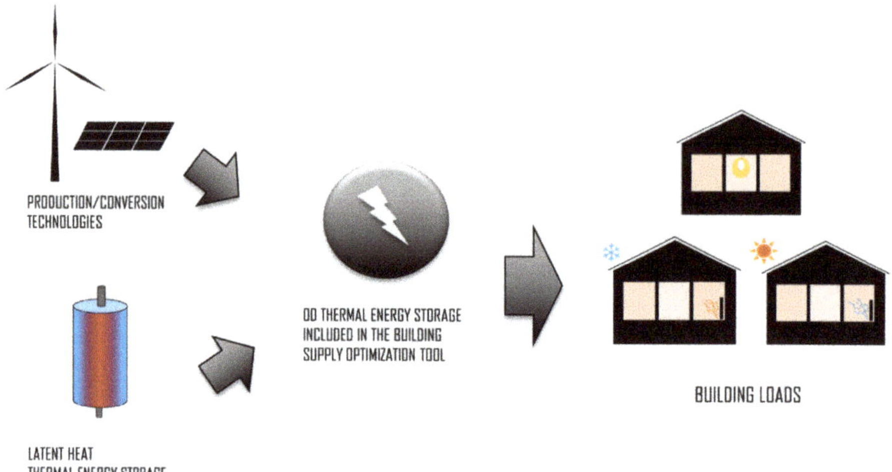

Figure 4. Schematics of the multi-energy system optimization with a 0D LHTS model.

4. Results and Discussion

In order to assess the validity of the 0D model, this section compares the results with the ones obtained from the 2D model. Afterwards, the impact of an LHTS system on the operational strategies of two relevant case studies is quantified through the application of the compact 0D model. In the former case, it shows how the LHTS allows achieving a significant reduction of the thermal power peak in a District Heating (DH) substation. In the latter case, it is used to include the latent thermal storage in an optimizer that finds the optimal performance of a multi-energy system.

4.1. Comparison between 0D and 2D Models

The coefficients resulting from the fitting procedure described in Section 2 are reported in Table 2. Equations (8) and (9) approximate very well the 2D power discharge and charge curves when the initial conditions are, respectively, $SOC_0 = 1$ and $SOC_0 = 0$ (Figures 5 and 6). The former fitting curve is characterized by a correlation coefficient $R^2 = 0.995$, while the latter has $R^2 = 0.992$. Overall, the minimum value is $R^2 = 0.939$, while the maximum standard deviation is 0.138 kW.

Table 2. Fitting coefficients in Equations (8) and (9).

Coefficient	Discharge Phase	Charge Phase
A	0.1752 [kW]	3.353 [kW]
B	3.112 [-]	−45.93 [-]
C	−0.2078 [kW]	1.337 [kW]
D	−0.9345 [-]	−3.606 [-]
K	1.758 [kW]	0.3296 [kW]
E	0.5518 [-]	0.5908 [-]
F	0.3442 [-]	0.4197 [-]

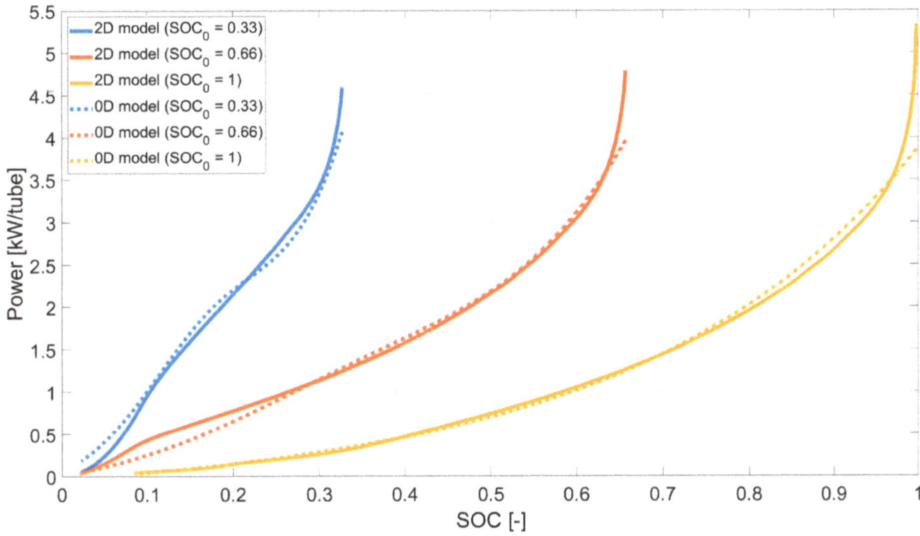

Figure 5. Latent Heat Thermal Storage (LHTS) discharge starting from different initial states of charge (after a partial/complete charge).

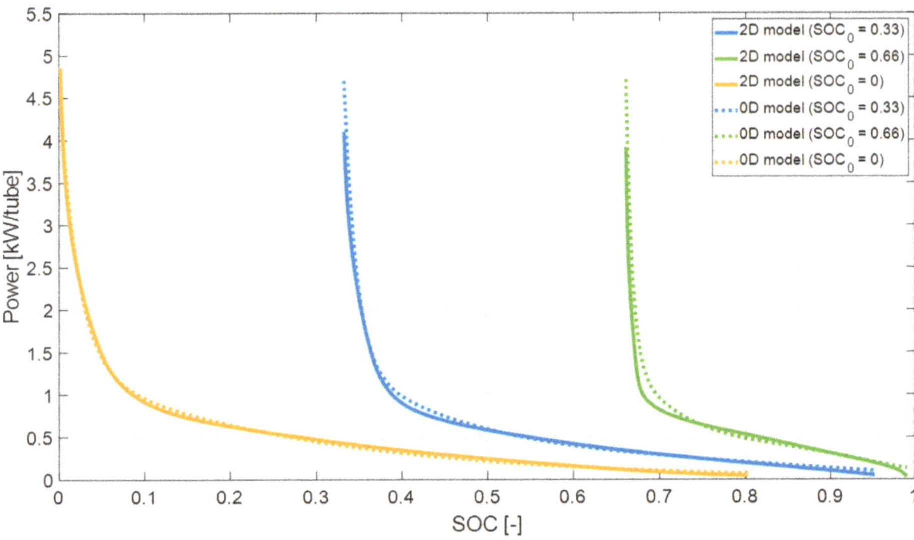

Figure 6. LHTS charge starting from different initial states of charge (after a partial/complete discharge).

The correspondence between the LHTS thermal power and its state of charge is not univocal. As a matter of fact, the LHTS thermal power depends both on the current state of charge (*SOC*) and on the initial state of charge (SOC_0). However, SOC_0 affects the results only in case a new charging or discharging phase starts. For instance, if there are two consecutive charging or discharging phases (only interrupted by an idle period), the value of SOC_0 should not be updated at the beginning of the second charge/discharge simulation. Instead, the previous value must be retained. This issue is due to the fact that when a partial charge is performed, only the PCM close to the fins becomes liquid, while the PCM far from the fins remains solid. If the subsequent operational phase is a discharge, a peak of thermal power occurs at the beginning of this latter process because the first layer of PCM encountered in the heat propagation is liquid (Figure 5).

A similar consideration is valid when a charge is performed, starting from a partial discharge of the system (Figure 6). The 0D model for the LHTS discharge thermal power (8) is slightly less accurate when SOC_0 is much smaller than 1 (i.e., when a discharge is preceded by a very short charging phase). However, it is rather unlikely to discharge a thermal storage starting from such a low content of energy. Similarly, the 0D model for the LHTS charge thermal power (9) is less accurate when the SOC_0 is much larger than 0 (i.e., when a charge is preceded by a very short discharge phase). As a matter of fact, the 2D curves in Figure 5 are both shifted and contracted with respect to one other, while the 2D curves in Figure 6 are mainly shifted. This dissimilarity might be due to the fact that the HTF inlet temperature during the charging phase (75 °C) is much closer to the PCM average phase change temperature (70 °C) compared to the HTF inlet temperature during discharge (48 °C). In fact, as indicated by [24], the HTF inlet temperature affects the LHTS thermal power.

4.2. Distributed LHTS in DH Networks

The 0D LHTS model is here applied to the reduction of the thermal power peak that is observed in DH networks at the beginning of their daily operation in some areas of the world (such as the Mediterranean area). In order to achieve this goal, the LHTS system follows the subsequent duty cycle, as detailed in Figure 7:

- The charging phase of each LHTS unit lasts 3 h; thus, the whole LHTS bundle is charged between 1.20 and 5.40 am (due to the imposed delay between the activation of each unit);
- The discharging phase of each LHTS unit lasts 1.5 h; thus, the whole LHTS bundle is discharged between 6.30 am and 9.20 am (due to the imposed delay between the activation of each unit).

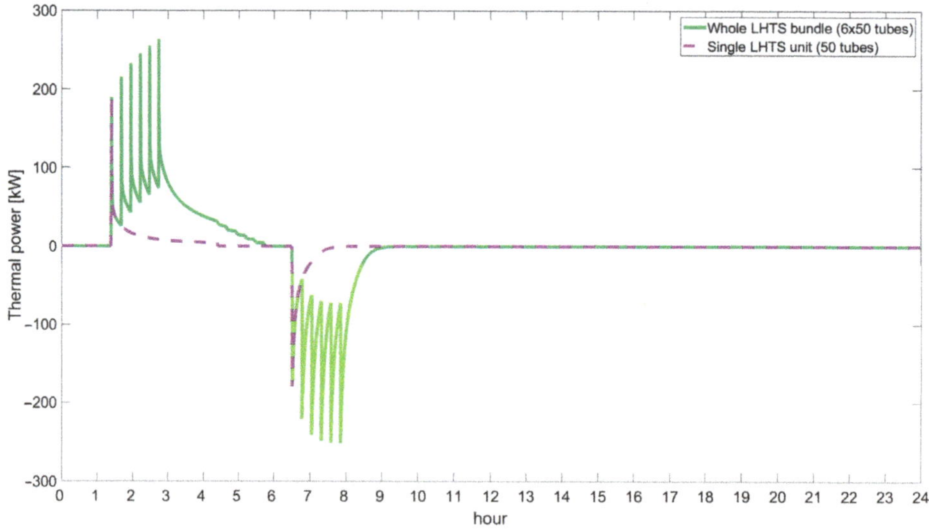

Figure 7. LHTS charge and discharge duty cycle to reduce DH morning peak.

Although all the energy absorbed by the LHTS during the charging phase is released in the discharging phase, the duration of these two processes is remarkably different. Since the HTF mass flow rate is constant, the thermal power is affected only by the temperature difference between the incoming HTF (48 °C or 75 °C) and the PCM average phase change temperature (70 °C) in both the operational phases. Consequently, the discharging phase is faster because the temperature difference is four times higher compared to the charging phase. Moreover, the spikes registered in the LHTS thermal power does not constitute a desirable feature from the point of view of the building heating system management. However, this highly unsteady behavior could be easily changed by a finer regulation of the HTF mass flow rate [16,17] (that is out of the goals of the present analysis). Furthermore, as detailed in Section 2.2, a perfect LHTS insulation is assumed. Consequently, heat losses are negligible when the storage is fully charged.

As shown in Figure 8, the time delay between the activation of each LHTS unit strongly affects the maximum power requested by the building to the DH network, whose reference value is 385 kW for this case study. Considering the LHTS system configuration analyzed, the peak thermal power of the DH supply is minimized when the activation of each LHTS unit is delayed by 16 min with respect to its preceding unit during the discharging phase. On the contrary, the minimization of the peak DH supply during the charge of the whole LHTS system would yield the obvious solution of charging each LHTS unit separately. However, this situation would not be compatible with the available time in the night. For this reason, an activation delay of 16 min was retained also for the charging phase.

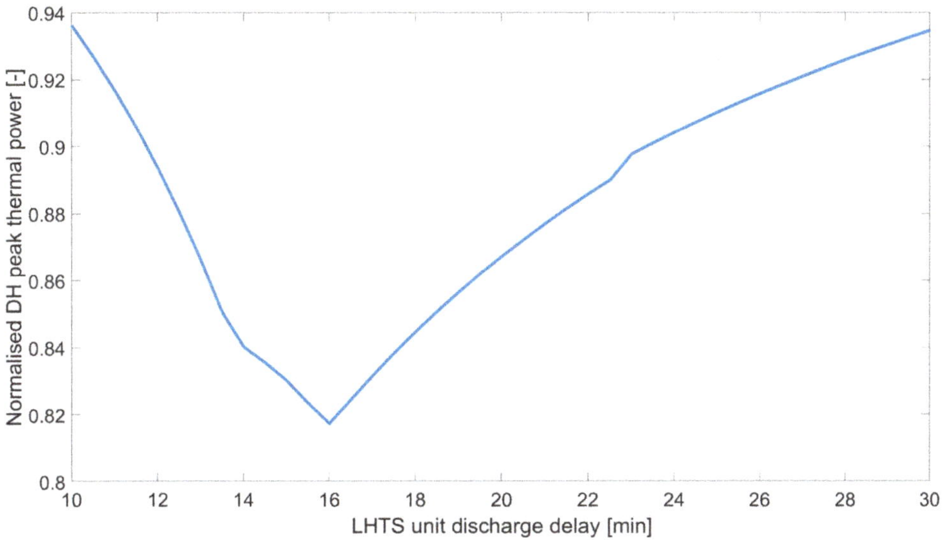

Figure 8. Minimization of DH peak thermal power through LHTS discharge delay.

Finally, Figure 9 shows the DH supply curve when the LHTS system presented in this study is deployed. For consistency with the data of the building energy demand, the DH supply curve describes the average thermal power over 15-min intervals. The morning peak reduction is evident from the figure. The thermal energy supplied by the DH between 6.30 and 9 am is reduced by 25.5%, while the maximum value of the DH thermal power is decreased by 18%. This result is achieved by simply shifting the supplied thermal energy in the night hours. This shifting could also be extensively advantageous for DH users in case of different tariffs during the daytime and nighttime. As an example, in Turin, the heat during the night is sold at a price that is up to 50% lower than the daily price [25]. Furthermore, the overall volume occupied by the LHTS system proposed in this study amounts to 3.4 m^3, while a sensible water storage tank would require 7 m^3 for the same energy content (assuming also the same temperature difference 75–48 °C). Therefore, this feature constitutes a great advantage when the thermal energy storage is located in the small technical rooms available in the buildings.

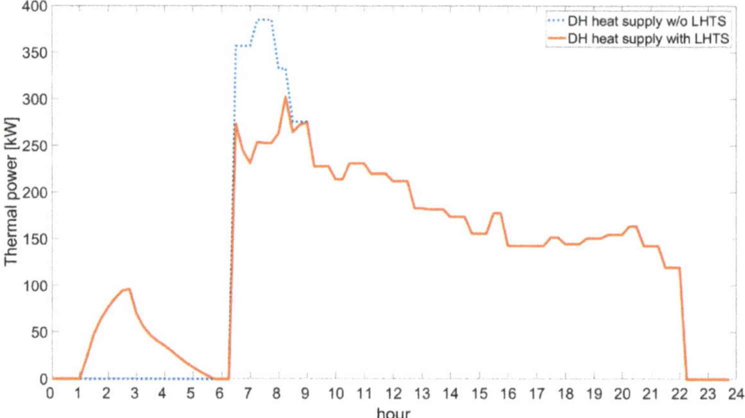

Figure 9. Building heat demand (15-min average thermal power) with and without LHTS.

4.3. LHTS in Multi-Energy Systems

This section reports the results achieved in the analysis of the Multi-Energy System. The technologies activated for supplying the electricity load are the micro-cogeneration (CHP) system and the photovoltaics (PV). The extra demand is supplied purchasing the electricity from the grid; the adoption of the batteries makes the production more flexible. The electricity consumption is due to both the building load and the electricity absorbed by the electric heat pump and the electric storage. Concerning the cooling load, this is supplied through an electric heat pump (EHP). As far as heating is concerned, Figure 10 includes the power evolution of the technologies selected by the optimizer. On the left of the figure, the production evolutions are reported; on the right, consumptions are reported. The production consists in the sum of the power due to the production/conversion technologies operated, the energy purchased from the grid, and the storage discharging phase. The consumption consists in the sum of the building load, the energy sold to the grid, the storage charging phase, and the consumption of the other technologies such as the heat consumed by the absorption heat pump. The sum of the production and consumption must be equal. As a result, the heat load is mainly provided by the micro-cogeneration system, the heat-only boiler, and the thermal solar. The LHTS allows storing energy in the valley to supply it during the peak loads.

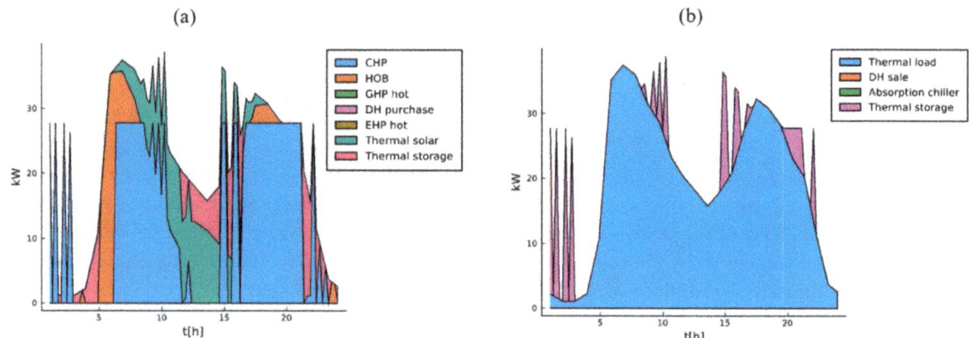

Figure 10. Production/consumption evolution for the energy vectors: (**a**) hot production; (**b**) hot consumption.

5. Conclusions

In this study, a simple 0D model for a modular shell-and-tube LHTS is presented. The model was obtained considering the main quantities affecting the charging/discharging evolution and through a fitting of the LHTS thermal power achieved with a more detailed CFD model. The compact model is obtained parameterizing the 2D model results as a function of the LHTS initial states of charge in order to account for partial charge and discharge. Results show that the 0D model is extremely accurate when compared with the 2D model, as the minimum value of the correlation coefficient (R^2) is 0.939, and the maximum standard deviation is 0.138 kW.

Overall, thanks to its simplicity and low computational cost, the 0D model can be easily used for simulations at the system level. It allows investigating the performance of numerous control strategies in energy systems equipped with an LHTS, even in case the LHTS is only partially charged or discharged. In this work, two specific relevant applications of the compact model are considered. The former regards the thermal power peak reduction in DH substations, while the latter concerns the optimal operational strategy of a set of production/conversion/storage technologies (included LHTS) in a multi-energy system. However, this study describes the functioning of the LHTS at the system level only on the basis of numerical results. Therefore, future work should test the LHTS also in a real case application to better quantify the model uncertainties.

Author Contributions: Conceptualization, A.C., E.G. and V.V.; methodology, A.C.; software, A.C. and G.M.; validation, A.C.; formal analysis, A.C.; investigation, A.C. and G.M.; resources, A.C.; data curation, A.C.; writing—original draft preparation, A.C. and G.M.; writing—review and editing, E.G. and A.L.; visualization, A.C. and G.M.; supervision, V.V., E.G. and A.L.; project administration, V.V.; funding acquisition, V.V. All authors have read and agreed to the published version of the manuscript.

Funding: This research was funded by EUROPEAN COMMISSION, grant number 815301. The APC was funded by EUROPEAN COMMISSION, grant number 815301.

Conflicts of Interest: The authors declare no conflict of interest.

References

1. Nguyen, T.-T.; Martin, V.; Malmquist, A.; Silva, C.A.S. A review on technology maturity of small scale energy storage technologies. *Renew. Energy Environ. Sustain.* **2017**, *2*, 36. [CrossRef]
2. Al-abidi, A.A.; Bin Mat, S.; Sopian, K.; Sulaiman, M.Y.; Mohammed, A.T. CFD applications for latent heat thermal energy storage: A review. *Renew. Sustain. Energy Rev.* **2013**, *20*, 353–363. [CrossRef]
3. Agyenim, F.; Hewitt, N.; Eames, P.; Smyth, M. A review of materials, heat transfer and phase change problem formulation for latent heat thermal energy storage systems (LHTESS). *Renew. Sustain. Energy Rev.* **2010**, *14*, 615–628. [CrossRef]
4. Niyas, H.; Prasad, S.; Muthukumar, P. Performance investigation of a lab–scale latent heat storage prototype—Numerical results. *Energy Convers. Manag.* **2017**, *135*, 188–199. [CrossRef]
5. Niyas, H.; Rao, C.R.C.; Muthukumar, P. Performance investigation of a lab-scale latent heat storage prototype—Experimental results. *Sol. Energy* **2017**, *155*, 971–984. [CrossRef]
6. Sciacovelli, A.; Gagliardi, F.; Verda, V. Maximization of performance of a PCM latent heat storage system with innovative fins. *Appl. Energy* **2015**, *137*, 707–715. [CrossRef]
7. Pizzolato, A.; Sharma, A.; Maute, K.; Sciacovelli, A.; Verda, V. Topology optimization for heat transfer enhancement in Latent Heat Thermal Energy Storage. *Int. J. Heat Mass Transf.* **2017**, *113*, 875–888. [CrossRef]
8. Pizzolato, A.; Sharma, A.; Maute, K.; Sciacovelli, A.; Verda, V. Design of effective fins for fast PCM melting and solidification in shell-and-tube latent heat thermal energy storage through topology optimization. *Appl. Energy* **2017**, *208*, 210–227. [CrossRef]
9. Esapour, M.; Hosseini, M.J.; Ranjbar, A.A.; Pahamli, Y.; Bahrampoury, R. Phase change in multi-tube heat exchangers. *Renew. Energy* **2016**, *85*, 1017–1025. [CrossRef]
10. Esapour, M.; Hosseini, M.J.; Ranjbar, A.A.; Bahrampoury, R. Numerical study on geometrical specifications and operational parameters of multi-tube heat storage systems. *Appl. Therm. Eng.* **2016**, *109*, 351–363. [CrossRef]
11. Seddegh, S.; Wang, X.; Henderson, A.D. A comparative study of thermal behaviour of a horizontal and vertical shell-and-tube energy storage using phase change materials. *Appl. Therm. Eng.* **2016**, *93*, 348–358. [CrossRef]

12. Neumann, H.; Palomba, V.; Frazzica, A.; Seiler, D.; Wittstadt, U.; Gschwander, S.; Restuccia, G. A simplified approach for modelling latent heat storages: Application and validation on two different fin-and-tubes heat exchangers. *Appl. Therm. Eng.* **2017**, *125*, 41–52. [CrossRef]
13. Parry, A.J.; Eames, P.C.; Agyenim, F.B. Modeling of Thermal Energy Storage Shell-and-Tube Heat Exchanger. *Heat Transf. Eng.* **2014**, *35*, 1–14. [CrossRef]
14. Tay, N.H.S.; Belusko, M.; Castell, A.; Cabeza, L.F.; Bruno, F. An effectiveness-NTU technique for characterising a finned tubes PCM system using a CFD model. *Appl. Energy* **2014**, *131*, 377–385. [CrossRef]
15. Johnson, M.; Vogel, J.; Hempel, M.; Hachmann, B.; Dengel, A. Design of high temperature thermal energy storage for high power levels. *Sustain. Cities Soc.* **2017**, *35*, 758–763. [CrossRef]
16. Colella, F.; Sciacovelli, A.; Verda, V. Numerical analysis of a medium scale latent energy storage unit for district heating systems. *Energy* **2012**, *45*, 397–406. [CrossRef]
17. Xu, T.; Humire, E.N.; Chiu, J.N.-W.; Sawalha, S. Numerical thermal performance investigation of a latent heat storage prototype toward effective use in residential heating systems. *Appl. Energy* **2020**, *278*, 115631. [CrossRef]
18. Vogel, J.; Johnson, M. Natural convection during melting in vertical finned tube latent thermal energy storage systems. *Appl. Energy* **2019**, *246*, 38–52. [CrossRef]
19. Agyenim, F.; Eames, P.; Smyth, M. Heat transfer enhancement in medium temperature thermal energy storage system using a multitube heat transfer array. *Renew. Energy* **2010**, *35*, 198–207. [CrossRef]
20. RT70HC Data Sheet. Available online: https://www.rubitherm.eu/media/products/datasheets/Techdata_-RT70HC_EN_06082018.PDF (accessed on 28 October 2020).
21. Ansys Fluent Theory Guide. Available online: https://ansyshelp.ansys.com/account/secured?returnurl=/Views/Secured/corp/v202/en/flu_th/flu_th_sec_melt_theory_energy.html%23flu_th_eq_melt_liq_frac (accessed on 28 October 2020).
22. Stamatiou, A.; Maranda, S.; Eckl, F.; Schuetz, P.; Fischer, L.; Worlitschek, J. Quasi-stationary modelling of solidification in a latent heat storage comprising a plain tube heat exchanger. *J. Energy Storage* **2018**, *20*, 551–559. [CrossRef]
23. Verda, V.; Colella, F. Primary energy savings through thermal storage in district heating networks. *Energy* **2011**, *36*, 4278–4286. [CrossRef]
24. Hosseini, M.J.; Rahimi, M.; Bahrampoury, R. Experimental and computational evolution of a shell and tube heat exchanger as a PCM thermal storage system. *Int. Commun. Heat Mass Transf.* **2014**, *50*, 128–136. [CrossRef]
25. District Heating Prices in Turin. Available online: https://www.irenlucegas.it/documents/66424/545179/Tariffe_TLR_nuovo_sito_1_10_2020.pdf/22923499-defa-4394-bc52-cf578e0602b6 (accessed on 28 October 2020).

Publisher's Note: MDPI stays neutral with regard to jurisdictional claims in published maps and institutional affiliations.

Article

Experimental Devices to Investigate the Long-Term Stability of Phase Change Materials under Application Conditions

Christoph Rathgeber [1,*], Stefan Hiebler [1], Rocío Bayón [2], Luisa F. Cabeza [3], Gabriel Zsembinszki [3], Gerald Englmair [4], Mark Dannemand [4], Gonzalo Diarce [5], Oliver Fellmann [6], Rebecca Ravotti [6], Dominic Groulx [7], Ali C. Kheirabadi [7], Stefan Gschwander [8], Stephan Höhlein [9], Andreas König-Haagen [9], Noé Beaupere [10,11] and Laurent Zalewski [10]

1 Bavarian Center for Applied Energy Research (ZAE Bayern), Walther-Meißner-Str. 6, 85748 Garching, Germany; stefan.hiebler@zae-bayern.de
2 Thermal Storage and Solar Fuels Unit, CIEMAT-PSA, Av. Complutense 40, 28040 Madrid, Spain; rocio.bayon@ciemat.es
3 GREiA Research Group, Universitat de Lleida, Pere de Cabrera s/n, 25001 Lleida, Spain; luisaf.cabeza@udl.cat (L.F.C.); gabriel.zsembinszki@udl.cat (G.Z.)
4 Department of Civil Engineering, Technical University of Denmark (DTU), Brovej, Building 118, 2800 Kongens Lyngby, Denmark; gereng@byg.dtu.dk (G.E.); markd@byg.dtu.dk (M.D.)
5 Faculty of Engineering of Bilbao, University of the Basque Country (UPV/EHU), Rafael Moreno Pitxitxi 2, 48012 Bilbao, Spain; gonzalo.diarce@ehu.eus
6 Institute of Mechanical Engineering and Energy Technology IME, Lucerne University of Applied Sciences and Arts (HSLU), Technikumstrasse 21, 6048 Horw, Switzerland; oliver.fellmann@hslu.ch (O.F.); rebecca.ravotti@hslu.ch (R.R.)
7 Lab of Applied Multiphase Thermal Engineering (LAMTE), Dalhousie University, 5269 Morris St., Halifax, NS B3H 4R2, Canada; dominic.groulx@dal.ca (D.G.); Dominic.Groulx@Dal.Ca (A.C.K.)
8 Fraunhofer Institute for Solar Energy Systems ISE, Heidenhofstr. 2, 79110 Freiburg, Germany; stefan.gschwander@ise.fraunhofer.de
9 Chair of Engineering Thermodynamics and Transport Processes (LTTT), Center of Energy Technology (ZET), University of Bayreuth, Universitätsstraße 30, 95447 Bayreuth, Germany; Stephan.Hoehlein@uni-bayreuth.de (S.H.); Andreas.koenig-haagen@uni-bayreuth.de (A.K.-H.)
10 Laboratoire de Génie Civil et géo-Environnement (LGCgE), ULR 4515, Univercity Artois, F-62400 Béthune, France; noe.beaupere@cea.fr (N.B.); laurent.zalewski@univ-artois.fr (L.Z.)
11 CEA/LITEN/DTNM/SA3D/LMCM, CEA Grenoble, Université Grenoble Alpes, 38000 Grenoble, France
* Correspondence: christoph.rathgeber@zae-bayern.de; Tel.: +49-893-294-4288

Received: 1 September 2020; Accepted: 29 October 2020; Published: 10 November 2020

Abstract: An important prerequisite to select a reliable phase change material (PCM) for thermal energy storage applications is to test it under application conditions. In the case of solid–liquid PCM, a large amount of thermal energy can be stored and released in a small temperature range around the solid–liquid phase transition. Therefore, to test the long-term stability of solid–liquid PCM, they are subjected to melting and solidification processes taking into account the conditions of the intended application. In this work, 18 experimental devices to investigate the long-term stability of PCM are presented. The experiments can be divided into thermal cycling stability tests, tests on PCM with stable supercooling, and tests on the stability of phase change slurries (PCS). In addition to these experiments, appropriate methods to investigate a possible degradation of the PCM are introduced. Considering the diversity of the investigated devices and the wide range of experimental parameters, further work toward a standardization of PCM stability testing is recommended.

Keywords: phase change materials (PCM); latent heat storage; degradation; thermal cycling stability; stable supercooling

1. Introduction

Phase change materials (PCM) can store large amounts of thermal energy in a small temperature range around their phase change temperature. Therefore, latent heat storage systems based on PCM can provide high thermal energy storage densities [1–4]. An important prerequisite to select a reliable phase change material (PCM) for thermal energy storage applications is to investigate its stability under application conditions. Since appropriate testing devices are not commercially available, lots of different self-built devices exist in various research institutions. Many of these institutions are currently participating in IEA ES Annex 33/SHC Task 58 "Material and Component Development for Compact Thermal Energy Storage", which is a joint working group of the "Energy Storage" (ES) and the "Solar Heating and Cooling" (SHC) Technology Collaboration Programmes (TCP) of the International Energy Agency (IEA) [5]. The main goal of this working group is to support an application-oriented development of innovative and compact thermal energy storage materials including PCM and thermochemical materials (TCM). Therefore, the participants have proposed sharing their experience regarding PCM stability testing. In total, 18 mostly self-built devices from 10 institutions are presented in this survey. The experiments performed can be divided in three types: type A: thermal cycling stability tests; type B: tests on PCM with stable supercooling, and type C: tests on the stability of phase change slurries (PCS).

In the case of solid–liquid PCM, heat is stored and released under repeated melting and solidification processes, which are also referred as thermal cycles. To assess the suitability of a PCM for a thermal storage application, testing the variation of its thermal properties under repeated thermal cycling (type A tests) taking application conditions into account is crucial [6–8]. Relevant experimental conditions are the temperature interval, heating and cooling rates, contact to atmosphere, contact to container, heat exchanger material, etc.

To study thermal cycling stability, the PCM under investigation is typically subjected to a temperature–time profile, as shown in Figure 1.

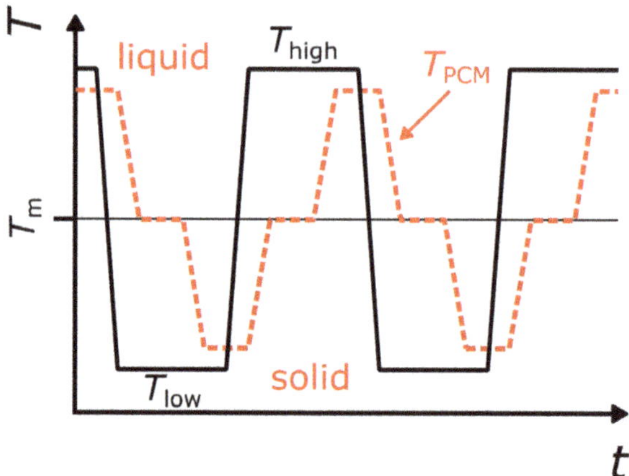

Figure 1. Idealized temperature–time curves of a thermal cycling stability experiment of a solid–liquid phase change material (PCM); PCM temperature (red dashed line); and heat transfer medium temperature (black solid line).

Two temperature levels, one above (T_{high}) and one below (T_{low}) the melting temperature (T_{m}), are applied alternately for a defined number of cycles. Certain periods of time are necessary to obtain a complete phase change from solid to liquid and vice versa. In between these two temperature levels, transition steps are carried out, either using temperature ramps with controlled temperature gradients, allowing natural cooling, or by heating and cooling as fast as possible. The curves plotted in Figure 1 are idealized. In reality (e.g., as shown in Figure 20), both the temperature–time profile of the heat transfer medium (solid black line) and the temperature response of the PCM under investigation (dashed red line) will not change suddenly between isothermal segments and transition segments. In addition, in many experiments, supercooling of the PCM will occur at least to a small extent.

Among the thermal properties to be checked after (or during) thermal cycling, the enthalpy–temperature curves upon melting and solidification are particularly relevant. In this context, a material can be regarded as stable after a certain number of thermal cycles if melting temperature and melting enthalpy variations remain within specific tolerance values. For example, according to the "RAL Quality Association PCM" [9] testing regulations, a decrease of the melting enthalpy by less than 10% and a melting temperature variation of ±1 K can be tolerated after a certain number of cycles. PCM are classified in cycling categories according to the number of cycles performed without damage: from A category (10,000 cycles) to F category (50 cycles). However, the testing regulations of the "RAL Quality Association PCM" neither indicate details on the appropriate design of stability testing devices nor give recommendations on suitable experimental techniques to study the stability of relevant properties after thermal cycling (except calorimetric techniques to measure enthalpy–temperature curves). In addition, apart from defining different cycling categories, nothing is said about the long-term stability of PCM under application conditions. So far, other testing regulations dedicated to the thermal cycling stability of PCM have not been proposed.

A concept for medium and long-term thermal energy storage is to utilize the ability of certain PCM to be stored as a liquid in the supercooled state. As investigated by Englmair et al. [10] and Desgrosseilliers [11], the application of sodium acetate trihydrate (SAT) in closed containers permits the use of its sensible heat capacity after melting while preserving its heat of fusion at room temperature in a state of stable supercooling. The result is a thermal energy storage capacity that can be used on demand by the controlled initialization of crystallization. This storage concept has been demonstrated for a solar heating system [12,13] and showed potential for compact thermal energy storage systems in buildings [14]. Recently, Duquesne et al. reported Xylitol [15] and Puupponen et al. reported micro-structured polyol-polystyrene composites [16] as further promising materials with stable supercooling properties. To investigate the stability of PCM with stable supercooling (type B tests), dedicated test procedures are required. In addition to temperature–time profiles of regular thermal cycling tests, both the degree and duration of supercooling are parameters of interest.

Phase change slurries (PCS) are fluids composed of PCM and a carrier fluid [1]. A PCS is stable if the emulsified or encapsulated PCM droplets do not degrade during operation, e.g., due to an agglomeration of droplets or a PCM leakage in the case of capsules. Among others, the stability of PCS depends on the investigated sample and particle size and the applied flow rate. Stability testing of PCS is carried out with self-built experimental devices (type C tests) that provide application-oriented charging and discharging conditions.

While existing studies reported on stability tests of a certain PCM tested with a certain device for a given number of thermal cycles (as e.g., in the review articles of Ferrer et al. [6] and Rathod and Banerjee [17]), this work focuses on the experimental devices and testing procedures that are being used to perform PCM stability tests under application conditions.

The aims of this work are as follows: first, to give an overview of the variety of experimental devices that are being used to investigate the long-term stability of PCM; second, to compare the specifications and test conditions of these devices, and to identify differences or similarities of the applied methods; third, to increase awareness that PCM stability studies need to take application conditions into account in order to investigate possible degradation phenomena in a realistic manner; and fourth, to initiate a discussion of whether there is a need to define or develop standardized stability test devices and procedures.

Following the comparison of investigated validation test devices of types A to C (Section 2), appropriate experimental methods to detect a possible degradation of the PCM are introduced (Section 3).

2. Comparison of Experimental Devices

2.1. Summary of Technical Specifications

This article is based on a survey comparing the experimental conditions and procedures of 18 testing devices from 10 different institutions. The device names used are formed according to the abbreviations of the respective institute designations. The technical specifications and the experimental conditions of these devices are summarized in Table 1. Devices are divided into thermal cycling stability tests (type A), tests on PCM utilizing stable supercooling (type B), and tests on the stability of PCS and encapsulated PCM (type C). The experimental devices for thermal cycling stability tests (type A) are typically used to test common PCM such as salt hydrates, paraffins, and other organic materials (e.g., fatty acids and esters). Further information about the investigated materials and their measured stability as well as further specifications of the devices can be found in the quoted references.

Table 1. Specifications of testing devices (device types A–C explained in Section 2.1). Explanations of columns: typical size (range) of the investigated sample; temperature range of measurements ΔT; heating and cooling rates between isothermal segments; typical number of melting–solidification cycles per day; atmosphere the PCM sample is in contact with; abbreviations used for stability tests: $T(t)_{cycles}$ = comparison of temperature–time curves of different cycles, $m(t)$ = sample mass monitoring after certain number of cycles or time, vis. = visually, Δh_{cycles} = comparison of the thermal energy content in repeated storage cycles, $T(t)_{solid.}$ = temperature profile during material solidification, Δp_{HX} = pressure loss of cooling heat exchanger, μ = viscosity measurement, further abbreviations in nomenclature; material of sample container.

Device Name (Type)	Sample Size	ΔT/°C	Heating, Cooling Rate/K·min^{-1}	Cycles Per Day	Atmosphere	Stability Tests	Sample Container	Ref.
CIEMAT I (A)	60–100 mL	RT [1]–350	1–20	1	air	$T(t)_{cycles}$, DSC, $m(t)$	glass, ceramic	[18,19]
CIEMAT II (A)	10 mL	RT–500	1–20, natural	1	air	$T(t)_{cycles}$, DSC, $m(t)$	ceramic	[18,19]
CIEMAT III (A)	60 mL	RT–500	1–20, natural	1	air, N_2, Ar	$T(t)_{cycles}$, DSC, $m(t)$	ceramic	[20]
EHU (A)	5 g	−45–200	variable	variable	air, N_2, Ar	DSC, vis., XRD, FTIR	glass, metal	-
HSLU (A)	1 × 100 mL, 3 × 25 mL, 5 × 8 mL	−40–180	10; 0.5	10	air	$T(t)_{cycles}$, DSC, vis.	glass	-
ISE capsules (A)	<22 L [3]	−10–80	1; 1	10	air	DSC, $m(t)$, RLM	polystyrene	[21]
ISE Peltier (A)	10–100 mL	−30–200	variable	<24	air	DSC, vis.	stainless steel	
LAMTE (A)	3–6 mL	−5–125	13; 7	<140	air	DSC	glass, plastic	[22–25]
LGCgE (A)	270 mL	5–80	0.05–0.3	2–3	air, vacuum	$T(t)_{cycles}$, Δh_{cycles}, $T(t)_{solid.}$	PMMA	[26]
LTTT (A)	10–100 ml	−20–180	variable	variable	air	$T(t)_{cycles}$, DSC	glass, metal, plastic	[27]
UDL-GREiA I (A)	18 × 0.5 mL	4–99	50; 50	100	air	DSC, FT-IR, TGA	polypropylene	[4,28]
UDL-GREiA II (A)	154 L	20–400	1; natural	1	air	DSC, FT-IR, TGA	stainless steel	[29–38]
UDL-GREiA III (A)	8 L	−10–80	0.5; 1	2	air	DSC, FT-IR, TGA	aluminum, stainless steel	[39]
ZAE (A)	60 ml	−30–220	1	2	air	$T(t)_{cycles}$, DSC, vis.	glass, stainless steel	[7]
DTU full-scale (B)	100–200 L	20–90	1; 0.5	0.2 (0.4 [2])	air	$T(t)_{cycles}$, Δh_{cycles}	metal, plastic	[40–42]
DTU heat loss (B)	200 g	20–90	1; natural	0.25 (0.5 [2])	air	$T(t)_{cycles}$, Δh_{cycles}, $T(t)_{solid.}$	glass jar, metal lid	[43]
DTU multiple (B)	10 × 30 L	8–93	0.15; 3–4	0.25	air	$T(t)_{cycles}$, Δh_{cycles}	stainless steel	[44]
ISE PCS (C)	3.5 L	−10–80	140; 100 [4]	1300 [4]	- [5]	DSC, Δp_{HX}, PSA, μ	stainless steel	[45]

[1] RT = room temperature, [2] without supercooling of SAT, [3] max. bath volume, [4] at 200 l·h^{-1}, [5] PCS in contact with water.

2.2. Descriptions of Testing Devices

In this section, the experimental setups, operating principles, and testing procedures of the self-built stability testing devices are presented. Devices are grouped and listed in the same order as in Table 1.

2.2.1. Type A—Thermal Cycling Stability Tests

CIEMAT I

The experimental device "CIEMAT I" is an oven that allows performing thermal tests under controlled heating and cooling rates with stand-by periods at constant temperatures up to 350 °C. Pictures of this device are shown in Figure 2. Thermal cycles can be performed between two temperature levels with transitions via linear ramps with rates from 1 to 20 K·min^{-1}. These ramps can be combined with stand-by periods up to a maximum test time of 24 h.

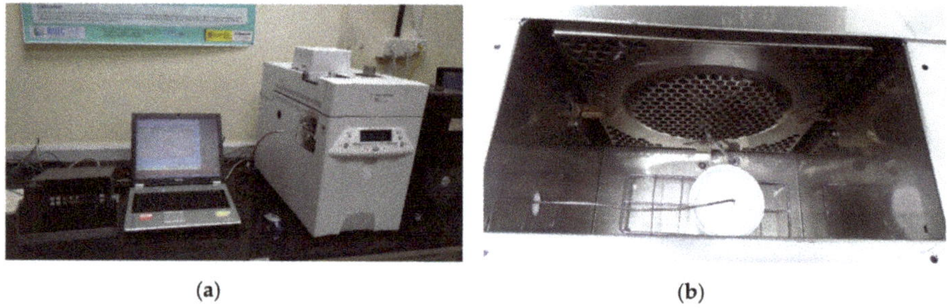

(a) (b)

Figure 2. Setup outside (**a**) and oven inside (**b**) of the device "CIEMAT I".

The temperature inside the oven is accurately controlled thanks to a fan located on one side, which provides forced ventilation (Figure 2b). Various samples with different sizes can be tested simultaneously under static ambient air. During the experiments, the temperature of both sample(s) and oven are monitored with calibrated T-type thermocouples.

CIEMAT II

The "CIEMAT II" device consists of a vertical oven with a cylindrical ceramic cavity as shown in Figure 3. This oven can perform thermal cycles up to 500 °C at controlled heating rates (between 1 and 20 K·min^{-1}) with stand-by periods at constant temperature. To cool down the sample, the heating system automatically switches off so that only natural cooling can be achieved. This device also allows performing thermal tests at constant temperature for indefinite periods of time. Moreover, it can be installed inside an extraction hood so that PCM thermal degradation tests under air can be carried out either at constant heating rate or constant temperature.

As shown in Figure 3b, only one sample can be tested at a time remaining under atmospheric air. During the experiments, the sample temperature is monitored with calibrated T-type or J-type thermocouples. Typically, one thermal cycle per day is carried out.

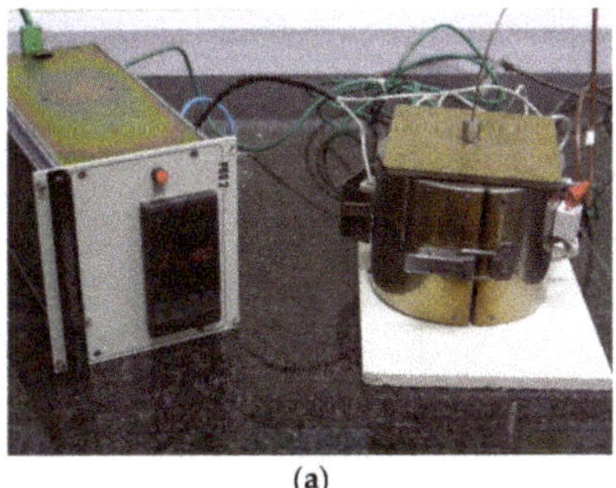
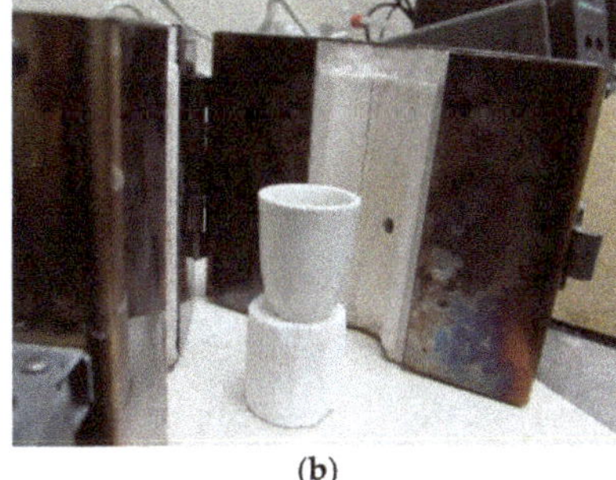

(a) (b)

Figure 3. Setup outside (**a**) and oven inside (**b**) of the device "CIEMAT II".

CIEMAT III

"CIEMAT III" is a thermal cycling device shown in Figure 4. It consists of a vessel enclosed inside a furnace that allows performing cycles under controlled heating (between 1 and 20 K·min^{-1}) up to 500 °C and natural cooling (Figure 4a) with stand-by periods in between. This device also allows performing thermal tests at constant temperature for indefinite periods of time. The testing vessel is a metallic cavity closed by two flanges where only one sample can be introduced. In the upper covering flange, a J-type thermocouple is installed to record the sample temperature during the experiments, and the lower flange has two connecting pipes for gas inlet and outlet, respectively (Figure 4b). Each gas pipe is connected to a valve that allows gas entrance and exit and keeps the pressure in the device at a user-defined pressure range. The gas facility is prepared for using Ar or N_2, but other gases such as air might also be used.

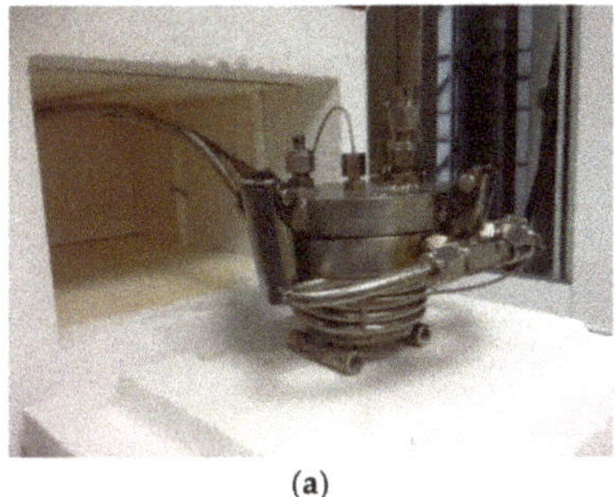

(a) (b)

Figure 4. Setup outside and furnace (**a**) and setup inside (**b**) of the device "CIEMAT III".

EHU

The testing device "EHU" is comprised of a thermostatic bath and airtight tubes. The samples are inserted inside the tubes, and these are immersed into the thermostatic bath, which uses silicone oil as a heat transfer fluid. The tubes are made of glass and include a nylon screwed cap with an O-ring sealing that ensures tightness. They can handle pressures up to 10 bar. The usual sample size is approximately 5 g, although tubes with variable dimensions can be employed (Figure 5).

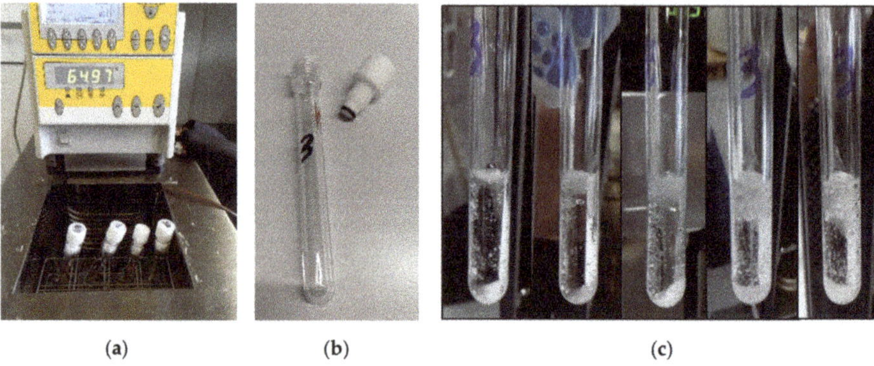

(a) **(b)** **(c)**

Figure 5. Pictures of stability testing device "EHU"; thermostatic bath (**a**), glass tube and cap (**b**), thermally cycled samples (**c**).

The arrangement and configuration are flexible and adapted to the investigated PCM and the kind of degradation expected. The temperature program usually comprises consecutive heating and cooling cycles if physical phase separation is expected. When the material under study might undergo thermal decomposition, then the program normally consists of submitting different samples to a constant temperature. The samples are extracted after different periods of time and analyzed. Several complementary techniques are used. Variations on the thermal behavior (storage capacity, melting temperatures, and others) are evaluated by differential scanning calorimetry (DSC). Visual control is kept by periodic imaging. Additional information is obtained by X-Ray diffraction, Fourier transform infrared spectroscopy (FT-IR), and liquid chromatography.

During the experiment, the temperature of the thermostatic bath is measured. The temperature inside of the samples can be also recorded by introducing a thermocouple inside a "sacrificed" sample. Normally, the PCM is in contact with air. However, other gases (e.g., N_2 or Ar) can be introduced by the use of a controlled atmosphere chamber (a globe box).

HSLU

The setup used at Lucerne University of Applied Sciences and Arts (HSLU) is based on the commercial device Easymax 102 (METTLER TOLEDO AG., CH), which is used as a cycling device and synthesis station (Figure 6). The Easymax is operated between −30 and 180 °C. Two heating chambers with various adapters allow the measurement with one to five vessels (1 × 100 mL, 3 × 25 mL, 5 × 8 mL) per chamber (Figure 7a). Magnetic or mechanical stirring can be applied. Easymax 102 contains one temperature sensor per chamber. Therefore, an external logging device to record the temperature of the other vessels is used.

Different heating and cooling rates can be applied, either using a fixed rate or by controlled heating so that a temperature difference between the sample temperature (T_r) and the temperature of the jacket (T_j) is not exceeded. If more than one vessel per chamber is measured, only heating by rate is possible (Figure 8).

During the experiments, the jacket temperature T_j and the reaction temperature T_r are measured. With the addition of a reaction calorimetry unit (Figure 7b), the phase change enthalpy, heat transfer coefficient, and specific heat can be determined as well. Typically, ten thermal cycles per day are carried out.

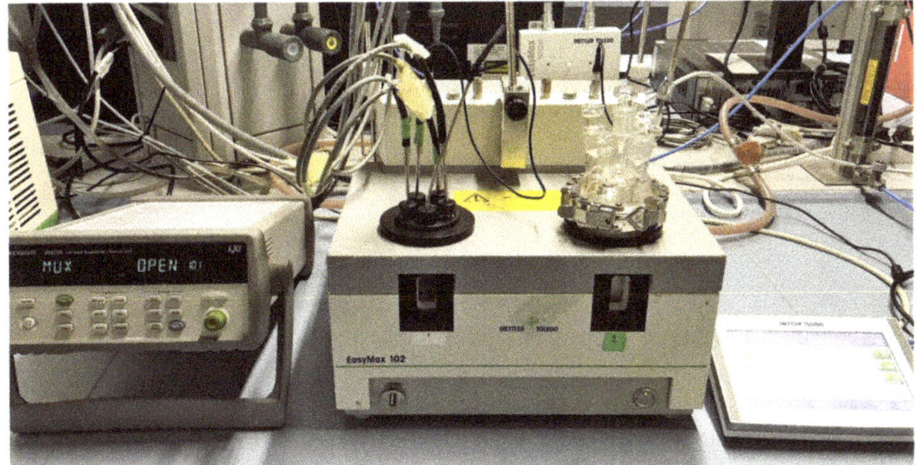

Figure 6. Picture of stability testing device "HSLU".

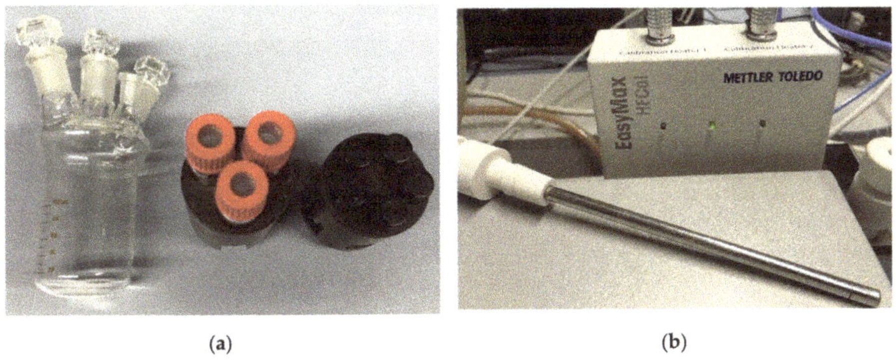

(**a**) (**b**)

Figure 7. Sample containers of the stability testing device "HSLU" (**a**), reaction calorimetry unit (**b**).

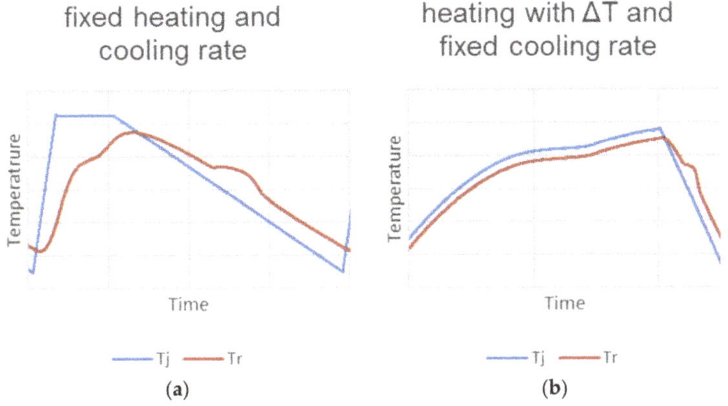

(**a**) (**b**)

Figure 8. Typical temperature profiles of the stability testing device "HSLU" (fixed heating and cooling rate (**a**), constant temperature difference upon heating and fixed cooling rate (**b**)); jacket temperature T_j, and reaction temperature T_r.

ISE Capsules

The test facility "ISE capsules" consists of an open bath in which the encapsulated PCM samples are immersed. Circulating tap water is applied as a heat transfer fluid. The flow rate can be adjusted manually by means of a pump with a downstream restriction. The bath temperature is controlled by a thermostat that supplies heat to the test circuit by a heat exchanger. Temperature sensors at the flow and return line of the bath and a volume flow sensor allow the thermal balancing of the heat stored in the PCM macro-capsules. The schematic of the setup is shown in Figure 9.

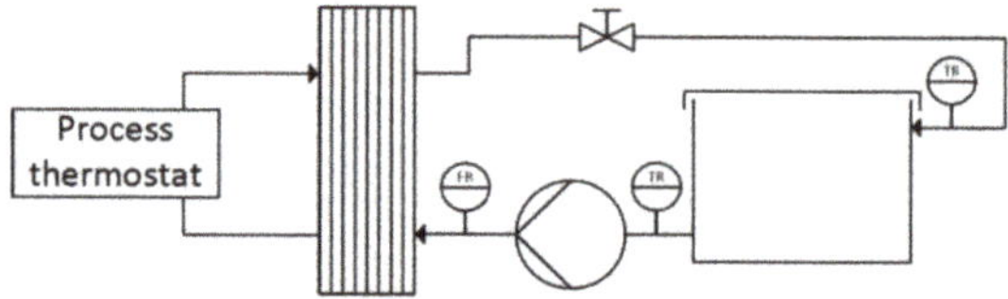

Figure 9. Schematic of stability testing device "ISE capsules".

The typical temperature profile is comprised of temperature steps (1 K·min^{-1} heating and cooling rate), which are repeated cyclically. The set temperatures depend on the PCM under investigation and are usually set 5 K below and above the melting temperature of the PCM. Typically, ten thermal cycles per day are carried out.

ISE Peltier

This is a flexible Peltier test rig for cycling in the temperature range −30 to 200 °C. Up to 16 Peltier modules with a maximum heating and cooling power of 30 W each and a surface of 65 × 65 mm can be flexibly combined in the x and z directions. Each Peltier module has two Peltier elements: one on the bottom and one on top. The upper Peltier elements are single point fixed to compensate for non-parallel surfaces. The contact pressure of the upper Peltier element can be controlled by pressurized air. A picture of the Peltier test rig is shown in Figure 10.

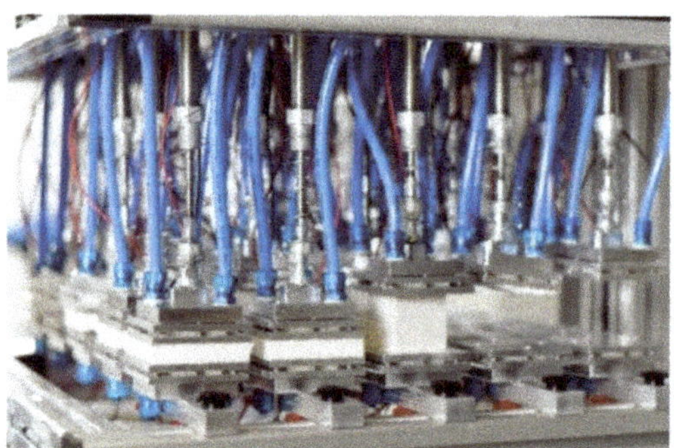

Figure 10. Peltier test rig.

Up to eight individual temperature controllers are used to control the temperature of samples. Peltier elements can be individually attached to each controller. Temperature profiles are composed of basic segments. Basic segments are the sudden temperature change, ramp, or isotherms. Heating and cooling rates can be defined individually. Sample crucibles are designed as needed. In line stability check is not foreseen. External stability check via DSC, optical inspection, or others is necessary.

Measured data are the temperatures of the PCM (middle or surface) and the temperature of the Peltier surface. Up to 24 thermal cycles per day are carried out.

LAMTE

The cycling device "LAMTE" is used to simultaneously cycle up to eight samples within a similar range of phase transition temperatures. The temperature is adjusted using two cartridge heaters embedded in an aluminum block as heat sources and four thermoelectric coolers for cooling of the system. Typically, a heating rate of approximately 13 K·min^{-1} and a cooling rate of approximately 7 K·min^{-1} are applied. Pictures of the device are given in Figure 11.

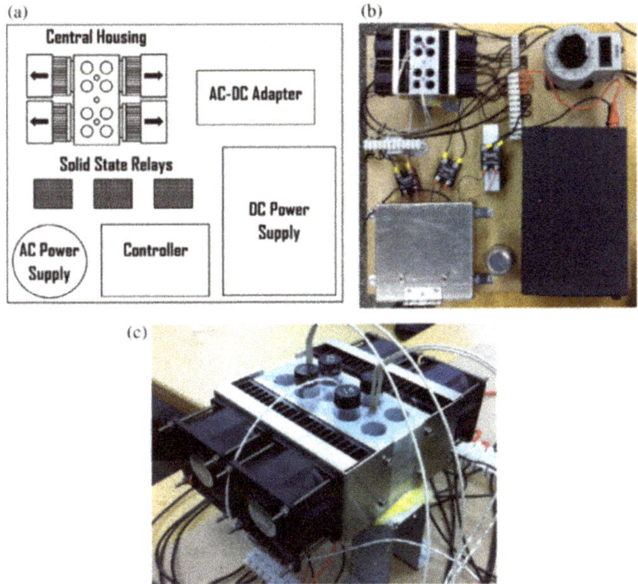

Figure 11. (**a**) Schematic diagram showing the various components of the thermal cycler general assembly, (**b**) photograph of the actual experimental setup, and (**c**) photograph of the housing unit with the heating and cooling elements and vials containing PCM (reproduced with permission from [22], Elsevier, 2016).

The maximum sample volume is 7.4 mL. Typically, only half the volume is used to allow for any expansion. The PCM is inserted in a dram vial made of borosilicate glass with a 15–425 threaded PTFE/silicone liner cap; the vial is inserted in the aluminum block where thermal grease is applied between the aluminum and the PCM vial to reduce thermal resistance to conduction. During the experiments, the temperature–time curve of the aluminum block is measured using a negative coefficient (NTC) thermistor. A typical temperature-time profile is shown in Figure 12. A numerical study of the system was done to determine the temperature, and phase, history of a PCM inside the dram vial during both the melting and solidification phase of the cycle, therefore ensuring complete melting and solidification of the PCM during each cycle [23]. Up to 140 thermal cycles per day are carried out.

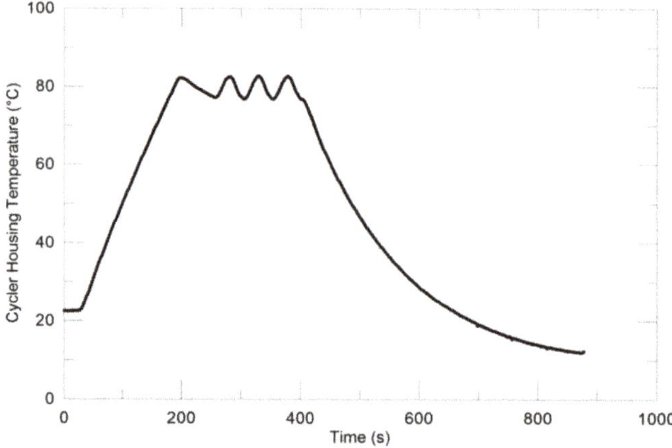

Figure 12. Typical temperature–time profile of stability testing device "LAMTE".

LGCgE

The Fluxmetric bench (Figure 13a) is a thermal cycling device used to impose temperature ramps on both larger sides of a parallelepipedal poly(methyl methacrylate) (PMMA) sample containing PCM (Figure 13b). Temperature ramps are applied thanks to the exchanger plates, each of which is connected to a refrigerated/heating circulator controlled by a computer. Typically, a heating rate of approximately 0.05 K·min^{-1} and a cooling rate of approximately 0.3 K·min^{-1} are applied. Two to three thermal cycles per day are carried out. The four smaller faces of the sample are insulated so as to minimize lateral heat losses (Figure 13b). In its current configuration, the container is filled with 270 mL of PCM. This configuration allows limiting convective flow when the PCM is in a liquid state. A heat flux meter incorporating a T-type thermocouple is placed on each side of the PMMA sample in order to calculate the energy balance at each heat cycle and to identify the melting and solidification temperatures. Between each temperature ramp, an isothermal step is imposed to return to an equilibrium (isothermal state and heat flux = 0) (Figure 14). For each fusion, the latent heat is estimated by subtracting the sensible energy stored (stored by PMMA and PCM) from the total stored energy measured by heat flux meters.

For example, multiple heating and cooling ramps have been imposed on SAT to study its aging by following the variation in the amount of latent heat involved in each melting transition [26]. At each cooling ramp, the temperature of the stochastic solidification (or degree of supercooling) can be identified. For each cooling ramp, by observing the peak of heat flux at the moment of crystallization, the degree of supercooling can be estimated (Figure 14). In Figure 14, part of a thermal cycling experiment on SAT is shown beginning with the crystallization of cycle 32 ("Crist. 32") up to cycle 35. Cycle 34 consist of melting (Melt. 34), cooling down without crystallization ("Sensible heat loss") followed by the crystallization of the supercooled liquid ("Crist. 34").

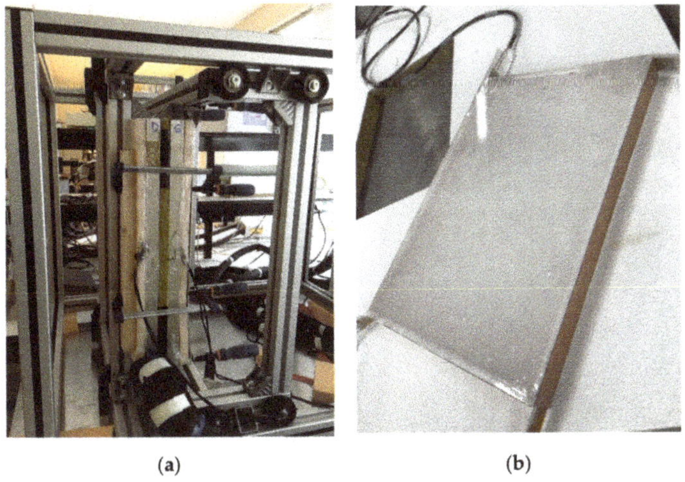

(a) (b)

Figure 13. Fluxmetric bench "LGCgE" (**a**), parallelepipedal poly (methyl methacrylate) (PMMA) sample containing PCM with the lateral heat fluxmeters before installation in the insulation (**b**).

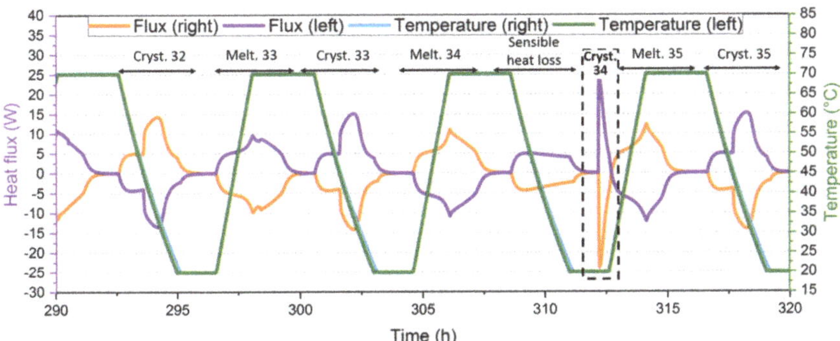

Figure 14. Ramps of temperature and heat fluxes measurements during thermal cycling of sodium acetate trihydrate (SAT).

LTTT

The thermal cycling device "LTTT" is comprised of a thermostatic bath (water or thermal oil) with a 4-channel PT100 data logger (expandable). Pictures of the device are shown in Figure 15.

As temperature profile, one temperature level above and one the below phase-change temperature with a rapid change between the temperature levels or with defined heating/cooling rates are applied. Up to 16 samples can be cycled simultaneously in test tubes (made of glass, metal, or plastic), which are sealed with a rubber plug and immersed in the thermostatic bath. During the experiments, temperature–time curves of the samples are measured. After a certain number of cycles, DSC measurements are performed, and the temperature curves of the samples are analyzed.

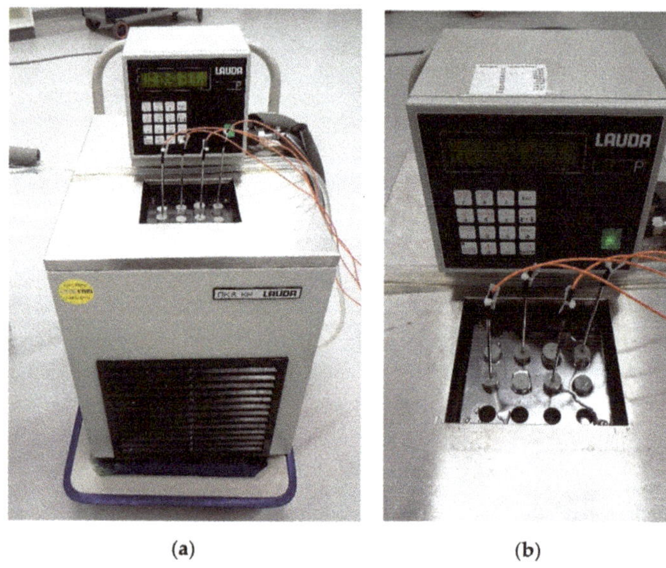

(a) (b)

Figure 15. View of the stability testing device "LTTT" (**a**) and detailed view of the arrangement of the sample holders (**b**).

UDL-GREiA I (This Equipment Is Shared with the University of Barcelona, DIOPMA Research Group)

The GeneQ BIOER TC-18/H(b) thermal cycler allows performing thermal cycles from room temperature up to 100 °C to investigate the stability of PCM samples over time. Typically, 50 K·min^{-1} is applied as the heating and cooling rate. Up to 100 thermal cycles per day are carried out. The thermal cycler allows placing 18 samples at the same time in 0.5 mL Eppendorf tubes. The Eppendorf tube must not be filled completely, as only the lower part will be subjected to heating and cooling cycles. It is recommended to fill 2/3 of the Eppendorf tube. Depending on the aim of the study, different temperature profiles can be applied. A picture of the stability testing device "GREiA I" is shown in Figure 16.

Figure 16. Picture of stability testing device "UDL-GREiA I".

UDL-GREiA II

The pilot plant at the University of Lleida was built to accurately test different thermal energy storage (TES) systems working with latent or sensible heat storage materials. This facility is composed of a 24 kW_e electrical boiler to heat up the heat transfer fluid (HTF) (acting as energy source during the charging process), different storage tanks containing PCM, and a 20 kW_e air heat exchanger to cool down the HTF (acting as energy consumption). Typically, one thermal cycle per day is carried out. A picture and schematic of the pilot plant test facility are shown in Figure 17.

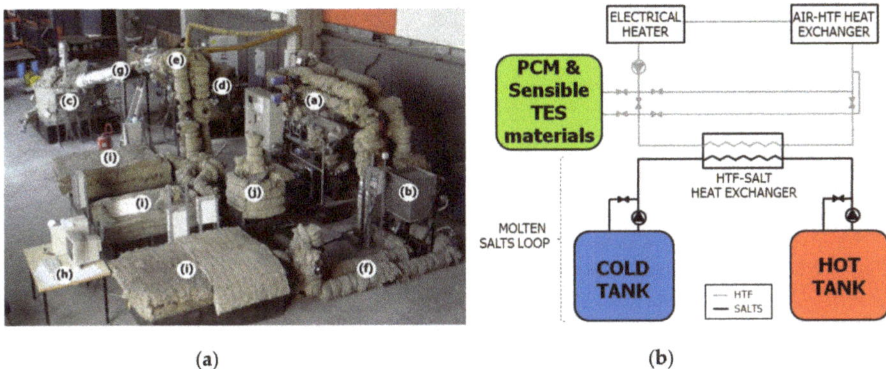

(a) (b)

Figure 17. (**a**) Picture and (**b**) scheme of stability testing device "UDL-GREiA II" (reproduced with permission from [46], University of Lleida, 2017). The letters in the figure (**a**) correspond to: (a) electrical heater, (b) air–heat transfer fluid (HTF) heat exchanger, (c) hot tank, (d) cold tank, (e) plate heat exchanger HTF–molten salt, (f) HTF loop, (g) molten salt, (h) acquisition and recording system, (i) latent thermal energy storage system, and (j) sensible thermal energy storage system.

A data acquisition system consisting of temperature, pressure, and flow rate sensors as well as different data loggers and a personal computer were integrated in the facility to measure the HTF flows, HTF pressures, and HTF and PCM temperatures in the boiler and storage tank, respectively. Depending on the aim of the study, different temperature profiles are applied. In this regard, during experimental tests that are carried out to assess the performance of components such as PCM tanks under different charging and discharging conditions, thermal cycling testing of the PCM can also be realized as far as the required number of thermal cycles and the temperature profile are known. Then, the degradation of PCM after different cycles can be evaluated at a pilot-plant scale, or small samples of the PCM can be extracted and analyzed in the lab using established techniques, such as DSC, FT-IR spectroscopy, or thermogravimetric analysis (TGA).

UDL-GREiA III

This experimental setup of the GREiA lab at the University of Lleida was built to test the behavior of different types of heat exchangers and thermal energy storage tanks. Both charging and discharging of the thermal energy storage modules can be performed by means of two separate circuits: one cooling circuit connected to a variable capacity condensing unit, and one heating circuit connected to a thermal bath. There is the possibility of connecting the heating bath to a water storage tank to increase the thermal inertia and to provide a more stable water temperature to the heat exchangers. A picture and schematic of the test facility are shown in Figure 18.

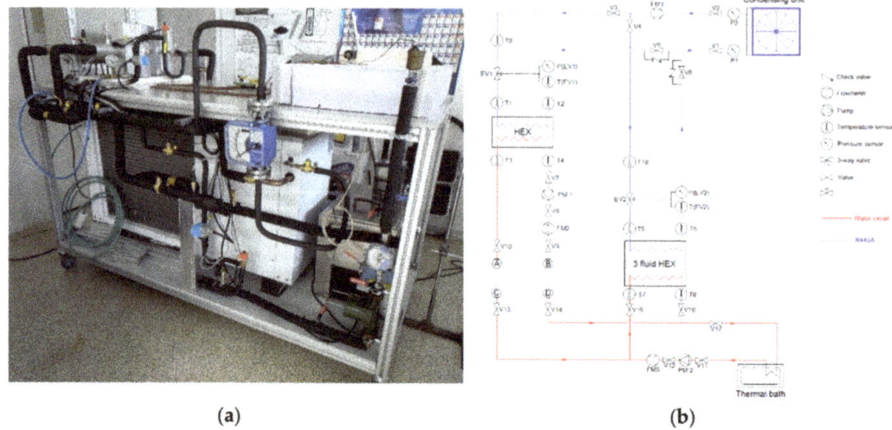

(**a**) (**b**)

Figure 18. Picture (**a**) and scheme (**b**) of test facility "UDL-GREiA III".

During the experiments, PCM and HTF temperatures, PCM pressure, and HTF flow rate are measured. Depending on the aim of the study, different temperature profiles can be applied. Due to the relatively small size of this setup, dedicated thermal cycling tests of PCM can be performed to study the degradation of the PCM after different numbers of melting–solidification cycles at lab scale. Typically, two thermal cycles per day are carried out.

ZAE

The thermal cycling test rig at ZAE Bayern can be applied to thermally cycle PCM under successive melting and crystallization processes and to visually detect the variation in appearance of the material, e.g., a phase separation. The test rig includes three sample containers to simultaneously investigate three PCM with similar melting temperatures [7]. A picture of the installation is given in Figure 19.

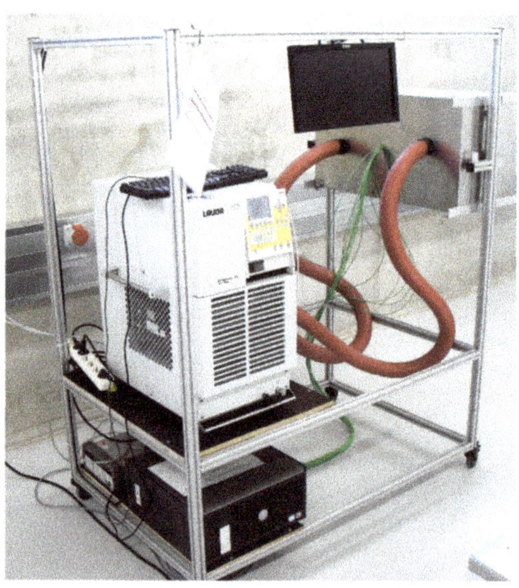

Figure 19. Pictures of stability testing device "ZAE".

To thermally cycle PCM, two different temperature levels for melting/freezing and transitions via linear ramps are applied (typically 1 K·min^{-1} heating and cooling rate). During the experiments, temperature–time curves of three temperature sensors within each of the three samples and three temperature sensors in the heat exchanger close to the sample containers are recorded. Figure 20 depicts the temperature–time data of one full thermal cycle.

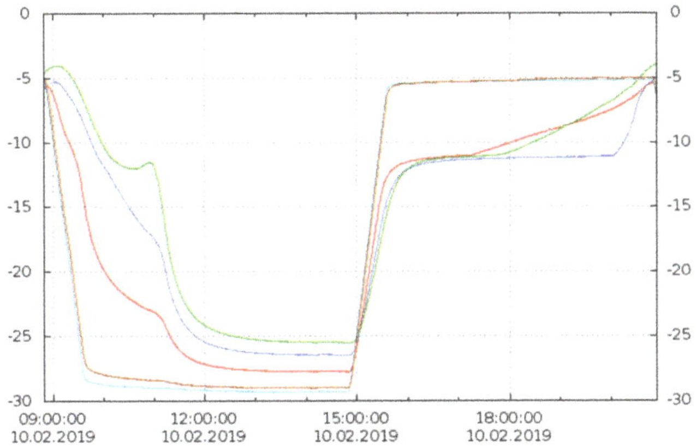

Figure 20. Temperature–time curves measured with the stability testing device "ZAE".

Usually, two thermal cycles per day are carried out. In the assessment of the cycling stability, special attention is paid to the visual detection of a possible phase separation. Figure 21 shows a thermally cycled sample at a temperature above the original melting point.

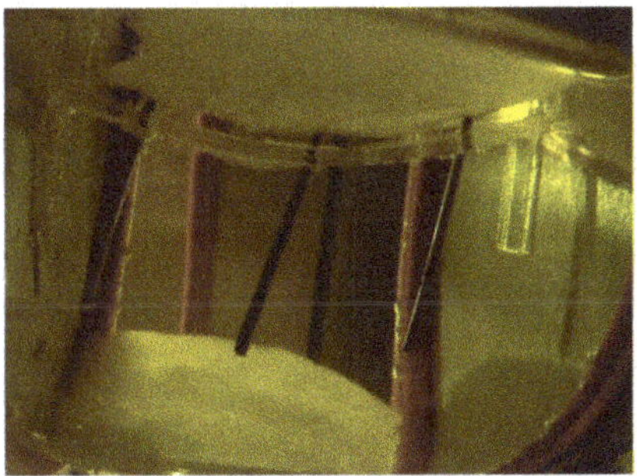

Figure 21. Visual observation of phase separation of a solid–liquid PCM in the device "ZAE".

The visual observation indicates that at least three different phases are present: a liquid and two solid phases, one at the bottom and one at the top of the sample container. If no phase separation is observed, after a certain number of cycles, a small liquid amount of sample material is taken and measured via DSC to check for potential variations of the melting enthalpy and temperature. Performing thermal cycling stability tests under application-relevant conditions and observing potential phase separation visually has shown to provide a reliable assessment of the long-term thermal cycling stability of PCM, e.g., of PCM based on salt-water mixtures.

Summary of Type A Tests

Summarizing the experimental devices to investigate the thermal cycling stability of PCM, a large variety in terms of applied test conditions can be observed. Investigated sample sizes range from

0.5 mL to 154 l. Heating and cooling rates vary from 0.05 to ≈50 K·min^{-1} and, as a consequence, the typical number of thermal cycles that are carried out per day varies from 1 up to 140. Different sample container materials are being used: metals, glass, ceramic, and plastics. The atmosphere the PCM sample is in contact with is in most cases air. Two devices allow the use of inert gases (N_2, Ar) instead of air; one device is capable of applying vacuum. In principle, a device that is used for thermal cycling tests can also be used to apply a constant temperature to the sample under investigation. For example, this applies to the devices CIEMAT I, CIEMAT II, and CIEMAT III. Overall, the observed variety in terms of applied experimental test conditions can be a consequence of strongly differing applications for which the stability tests have been performed and/or an indication for a lack of standardization regarding PCM thermal cycling stability testing.

2.2.2. Type B—Tests on Supercooled PCM

In the case of PCM with stable supercooling, the investigation of long-term stability consists mainly in determining the reliability of the state of supercooling over a certain period of time under various experimental conditions (size of the sample container or storage tank, maximum charging temperature, minimum discharging temperature, etc.). The desired stability of a supercooled PCM storage unit is achieved if no spontaneous crystallization occurs during the discharge phase of the storage unit or during the holding time of the long-term storage phase. Since PCM storage systems based on stably supercooling PCM are usually operated as long-term storage systems, the number of thermal cycles to be investigated is comparatively low.

DTU Full-Scale

"DTU full-scale" is a full-scale heat storage prototype testing installation [40–42]. The stability of the supercooled PCM in contact with the container material is investigated. The setup is shown in Figure 22.

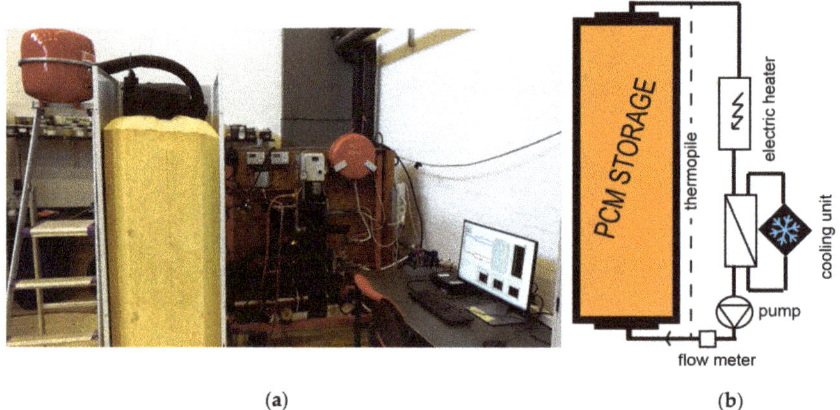

Figure 22. Testing device "DTU full scale": PCM heat storage test facility at DTU (**a**), schematic of charge and discharge loop (**b**).

A typical temperature profile applied to study the supercooling of SAT consists in the following steps (Figure 23): charging from approximately 20 to 90 °C, cooling down to ambient temperature (approximately 20 °C), initialization of crystallization, and discharge to ambient temperature (approximately 20 °C).

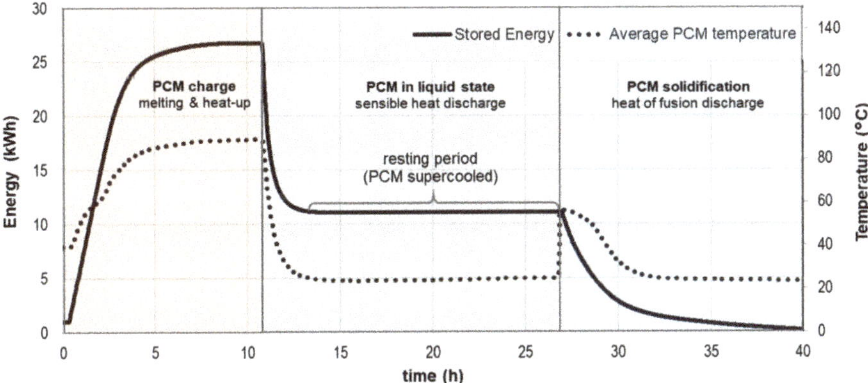

Figure 23. Temperature profile of a test cycle with supercooling of PCM (device "DTU full scale").

During the experiments, the following data are measured: temperatures (inlet, outlet) and volume flow rate of the heat transfer fluid, ambient temperature, and heat storage surface temperature.

DTU Heat Loss Method

With "DTU heat loss method", determination of the heat content of (supercooled) PCM at a large sample size (approximately 200 g) is performed [43]. Pictures of the testing device are shown in Figure 24.

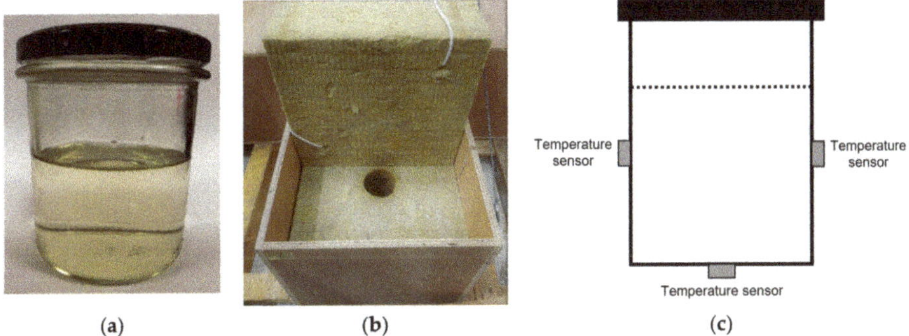

Figure 24. (**a**) Glass jar with lid, (**b**) the well-insulated box, (**c**) locations of temperature sensors ("DTU heat loss method").

In the case of supercooled SAT, a typical temperature profile consists in heating up the sample in the oven from approximately 20 to 90 °C followed by cooling down to ambient temperature at approximately 20 °C. Then, the crystallization of supercooled SAT is initiated inside an insulated chamber by means of a seed crystal, and the temperature–time curve during the temperature rises due to the crystallization enthalpy, and the succeeding cooling down to ambient temperature is measured. Figure 25 depicts the temperature development of a SAT sample, where the heat content was evaluated as the thermal energy dispersed to the ambience (green area).

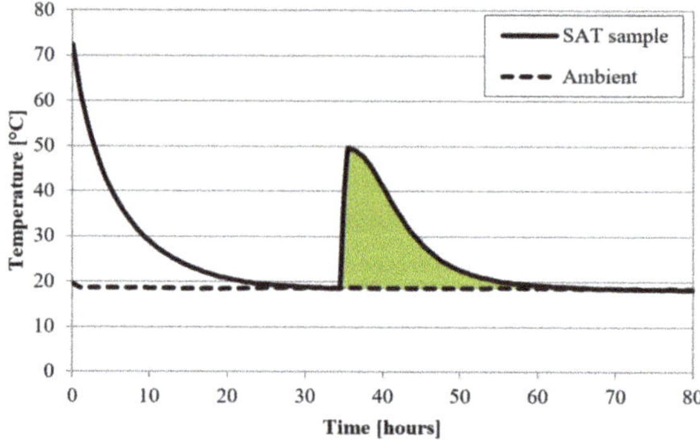

Figure 25. Exemplary cooling process of PCM samples in the device "DTU heat loss method".

DTU Multiple Storage

The device "DTU multiple storage" consists of multiple (10) identical heat storage units [47] which are subjected to repeated heating and cooling cycles with various temperature levels and durations [44]. SAT has been investigated as PCM utilizing supercooling. The focus of the investigations is the effect of different temperature levels and duration of charge on the supercooling stability of SAT. Illustrations of the testing device are given in Figure 26.

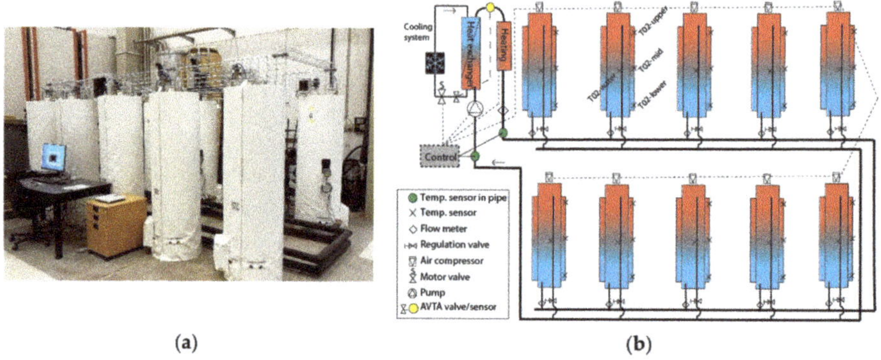

Figure 26. Testing device "DTU multiple storage": PCM heat storage test facility at DTU (**a**), schematic of charge and discharge loop (**b**).

Regarding application conditions, discharge is conducted with cold water at 10 °C. As the focus of investigations is the supercooling behavior of SAT, a typical temperature profile consists of the following steps: charging in periods of 8 to 16 h from 10 to 90 °C, cooling down to 10 to 30 °C, 72 h observation period (looking for spontaneous crystallization of the supercooled SAT), intentional initialization of solidification and discharge (at 10 to 30 °C). During the experiments, the following data are measured: temperatures (inlet, outlet) and volume flow rate of the heat transfer fluid, ambient temperature, and heat storage surface temperature.

Summary of Type B Tests

In this study, the experimental setups to investigate the stability of PCM with stable supercooling are dedicated to the properties of SAT. Due to the melting temperature of SAT at 58 °C and its significant supercooling, the experimental devices in this section are being operated within the temperature range of approximately 10 to 90 °C. As the three experimental devices in this section have been designed by the same research group, it is not surprising that they cover different parts of stability investigations on supercooled SAT. The "DTU heat loss method" was applied to identify promising SAT composites by screening samples of up to 200 g with a variety of additive concentrations. The "DTU full scale" setup was used to test prototype stores of up to 200 l storage volume with previously identified promising SAT composites in combination with container and heat exchangers, which are designed to enable stable supercooling. The test setup "DTU multiple storage" was designed to perform long-term stability tests. The operation of up to 10 storage prototype units with 30 l storage volume with typical application conditions (under variation of charging temperature, charging duration, and supercooling periods) is used to elucidate the reliability of developed heat stores.

2.2.3. Type C—Tests on Phase Change Slurries (PCS)

ISE PCS

The device "ISE PCS" (Figure 27) is a phase change slurry cycling test rig to determine the stability of PCS while cycling through a hydraulic system in which the dispersed PCM is exposed to repeated phase transitions and shear stress caused by a pump and other hydraulic components. It consists of three hydraulic circuits: the PCS test circuit, a heating circuit, and a cooling circuit; the last two are connected to thermostats (Figure 28). The heat flux in these two circuits is recorded by measuring the volume flows and the temperature difference over the heat exchangers. In the PCS circuit, the volume flow and the temperatures over the heat exchangers are measured as well, so it is possible to monitor the heat flux into and out of the PCS. Additionally, the pressure difference over the cooling heat exchanger on the PCS side is recorded. Samples of PCS for analysis can be taken while the test facility is in operation.

To cycle PCS, the following temperature profile is typically applied: constant temperature at the output of the heating and cooling heat exchanger in the secondary circuit with temperature levels depending on the phase transition temperature range of the dispersed PCM. The cycle rate is determined by the volume of PCS filled into the secondary circuit and the volume flow applied. At 200 l·h^{-1} volume flow, approximately 1300 cycles per day are carried out. During the experiments, the following data are measured: temperatures at the inlets and outlets of both heat exchangers in the primary and secondary circuits; volume flows in all three circuits; and pressure loss in the secondary circuit of the cooling heat exchanger. The detection of degradation is done on five different parameters: (a) heat flux into and out of the PCS, (b) pressure drop over the cooling heat exchanger, (c) particle size distribution, (d) enthalpy determination via DSC, and (e) viscosity measurement of samples taken.

Figure 27. Picture of stability testing device "ISE PCS".

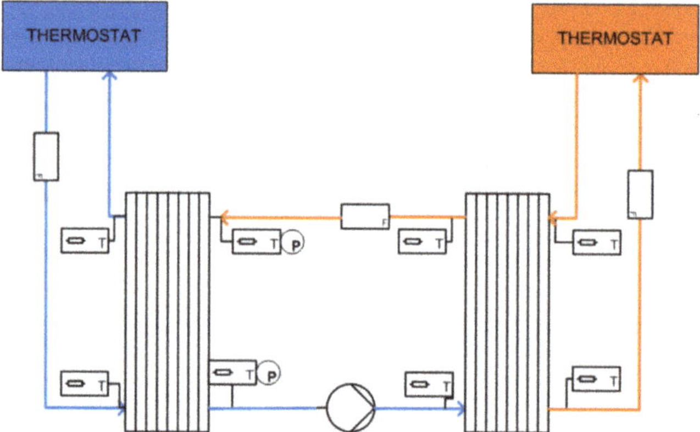

Figure 28. Schematic of stability testing device "ISE PCS".

Summary of Type C Tests

For this survey, one device for the investigation of the stability of PCS has been considered. Comparing the PCS test rig with the other stability testing devices of type A and type B, it can be observed that the PCS test rig applies the highest heating and cooling rates and the highest number of thermal cycles that are carried out per day to its samples. Such a fast melting and solidification of the sample is possible because PCS consist of PCM in comparably small volumes (emulsified or encapsulated) leading to a large heat transfer area and a relatively fast phase change between solid and liquid. Due to the special properties of PCS, the methods for investigating possible degradation processes also differ from those used for PCM and PCM with stable supercooling.

3. Experimental Techniques to Check a Possible Degradation of PCM

In this section, appropriate methods to check for a possible degradation of the PCM after testing under application conditions are introduced. These methods include an analysis of temperature–time curves (Sections 3.1 and 3.2), calorimetric measurements and measurements of thermal energy content (Sections 3.3, 3.5 and 3.6), mass change studies (Sections 3.4 and 3.5), structural investigations (Section 3.7, Section 3.8, Section 3.9, Section 3.10) and tests dedicated to PCS (Section 3.11, Section 3.12, Section 3.13). The abbreviations of the methods used in Table 1 are indicated in parentheses in each subsection title. Information on the uncertainties of standard measuring instruments (e.g., DSC) is discussed in the respective publications on these measuring methods. Uncertainties of special methods that are not based on standard measurements (e.g., Sections 3.1, 3.2, 3.6 and 3.7) have to be determined in each individual case and cannot be discussed adequately in the context of this paper.

3.1. Comparison of Temperature–Time Curves of Different Cycles ("$T(t)_{cycles}$")

The shape of temperature–time curves upon melting and solidification of a PCM indicate the temperature range of solid–liquid phase change. In addition, the degree of supercooling can be determined by evaluating the temperature–time curves upon cooling. The idea of comparing temperature–time curves of different cycles performed during a thermal cycling test measurement is based on the assumption that a possible degradation of the PCM will have an influence on its thermal properties and, hence, will affect the shape of temperature–time curves during the phase change.

In the case of stability tests on PCM utilizing stable supercooling (type B), temperature–time curves indicate if and when unwanted crystallization occurs.

3.2. Temperature Profile During Material Solidification ("$T(t)_{solid}$")

Knowing the heat loss coefficient of a vessel containing supercooled PCM in temperature equilibrium with the ambience, the heat content can be evaluated as the thermal energy dispersed to the ambience. This method was applied for stability tests on PCM utilizing stable supercooling (type B). If the heat content measured in successive storage cycles is reduced, degradation of the PCM may have occurred. A change in the plateau temperature during solidification can also indicate a degradation of the PCM.

3.3. Differential Scanning Calorimetry ("DSC")

Differential scanning calorimetry (DSC) is a widely used technique to measure the melting temperature and melting enthalpy of PCM. With DSC measurements, enthalpy–temperature curves upon heating and cooling can be determined [48]. Due to the relatively small sample size (in the order of 5 to 20 mg), the degree of supercooling observed in DSC is increased compared with the behavior of PCM in larger quantities [49,50]. Therefore, the analysis of DSC heating curves is more favorable in the context of PCM stability testing: DSC measurements are applied to check if the melting temperature and enthalpy (or the enthalpy–temperature curves) after the testing under application conditions (e.g., thermal cycling tests) are the same as before the test.

3.4. Sample Mass Monitoring after Certain Number of Cycles or Time ("m(t)")

A clear indication that a certain PCM undergoes some kind of degradation is the decrease of mass with time when it is either cycled or kept at constant temperature above the melting point [51]. Mass decrease can be due to vaporization and, hence, to material loss but also to the formation of volatile species due to the occurrence of chemical reactions so that the PCM composition changes. If mass loss is due to chemical reactions leading to volatile species, not only PCM thermal properties are expected to change but dangerous compounds can be evolved [8]. Note that if PCM is reacting with the surrounding gases (mainly water vapor or oxygen), its mass could increase as well. Therefore, if mass variation is observed for a PCM, the material cannot be implemented in any storage system unless measures are taken for avoiding such behavior (encapsulation, inert environment under pressure, etc.). Ensuring that PCM mass does not change over time under application conditions is a key issue for its validation. Simple experimental procedures such sample weighting after a certain number of cycles or time interval [51] can be designed for detecting the occurrence of possible PCM volatilization and/or degradation.

3.5. Thermogravimetric Analysis ("TGA")

Another way to study PCM mass variation and possible degradation reactions is performing thermogravimetric analysis (TGA). TGA results can be compared with the results of sample mass monitoring to obtain more information about PCM stability. TGA is a technique that monitors the weight of a sample within a furnace to study the mass changes within a temperature range. This technique is suitable for different thermal analysis, such as decomposition temperature, oxidation, or loss of volatiles. A material sample of a few mg is placed inside the TGA apparatus usually exposed to air or nitrogen atmosphere, which is able to heat it at a controlled rate or maintain it at a defined temperature. The mass of the sample is recorded and can be plotted against temperature or time by means of a thermogravimetric curve (thermogram).

In the case of PCM, TGA can be used to determine its maximum working temperature, usually defined as the temperature at which the sample loosed 1.5 wt % of its initial mass or after the dehydration process, and also the final degradation temperature, which can be extracted from the thermogram. Moreover, if TGA measurements are performed at various heating rates, degradation kinetics could be obtained as well and hence an estimation of the PCM lifetime [52].

3.6. Comparison of the Thermal Energy Content in Repeated Storage Cycles ("Δh_{cycles}")

The thermal energy content of investigated samples determined for repeated storage cycles indicates cyclic material stability for applied test conditions (sample size, temperatures, cooling and heating rates). Observations can be analyzed regarding the initial storage cycle and/or in comparison with different material compositions (e.g., for the identification of the optimal amount of additives). Furthermore, the dependency of cyclic stability on the degree and the duration of PCM supercooling (type B) can be investigated. The thermal energy content of prototype storages including PCM, heat transfer fluid, and container material determined under typical application conditions is used to check the reliability of the investigated storage.

3.7. Visually ("vis.")

In the case of solid–liquid PCM, a visual observation of possible phase separation is a useful and simple measure to detect a degradation of a PCM during experimental testing. The presence of solid material at a temperature where the PCM should be completely molten is a clear indication of phase separation. Thereby, a visual control of phase separation can be favorable compared to measured temperature–time curves. If the temperature sensor is not in thermal contact with the precipitated solid, the measured temperature–time curves may not be affected by the phase separation. Practically, a visual check of degradation can be carried out by taking pictures at given time intervals or temperature steps followed by an analysis of pictures from different thermal cycles. In addition, simply looking at the sample after the experimental investigations is a reliable measure to visually check degradation.

In the case of degradation tests performed on organic PCM at elevated temperatures, a visual check also includes a possible change in the PCM color. For example, a browning of sugar alcohols indicates a thermal degradation of the sample [53].

3.8. X-ray Powder Diffraction ("XRD")

X-ray powder diffraction (XRD) is mainly employed for the phase identification of a crystalline material, which can be used to perform compound identification. Therefore, the material is finely ground, homogenized, and the average bulk composition is determined. Based on these features, XRD can be easily applied to determine changes in the crystalline structure or the formation of new materials of a solid PCM due to a degradation process.

XRD measurements are presented as a diffractogram, where the X-ray intensity is plotted as a function of the scanning angle. The typical aspect of a diffractogram consists of a horizontal baseline alternated with sharp peaks denoted as diffractions. These diffractions are used to identify the compounds present in the sample. If, e.g., due to a degradation on thermal cycling, new solid compounds appear in the solid PCM, they can be identified when comparing the diffractogram of the original sample with that of the degraded sample. Mass concentrations as low as 2% with respect to the sample mass can be detected.

A potential phase separation can also be investigated by extracting aliquots from different locations of the PCM sample submitted to thermal cycling and comparing their phase/compound compositions by XRD. Some XRD devices can perform measurements under different temperatures, giving the possibility to study the variations of the structure of the solid PCM upon heating or cooling.

3.9. Fourier Transform Infrared Spectroscopy (FT-IR)

Fourier transform infrared (FT-IR) spectroscopy is an analytical technique based on the vibrations of a molecule that is irradiated with infrared radiation at a specific wavelength range due to selective absorption of infrared radiation by the sample material. Since the wavelengths that are absorbed by the sample material are characteristic of its molecular structure, this technique allows identifying molecular components and structures of organic and inorganic materials.

FT-IR with attenuated total reflectance (ATR) allows analyzing the PCM chemical degradation caused by thermal cycling. The use of ATR has the advantage that the sample does not require any previous specific preparation to obtain the spectra. When the material is oxidized or degraded as a result of thermal cycling, a total or partial disappearance of the characteristic peaks and/or the appearance of new peaks are observed in the spectra.

3.10. *Reflected Light Microscopy (RLM)*

In the case of the "ISE capsules" test rig, reflected light microscopy (RLM) is used for repeated optical inspections of PCM macrocapsules. With RLM, possible damage of the capsules walls can be detected that would lead to leakage of the PCM out of the capsules.

3.11. *Pressure Loss of Cooling Heat Exchanger (Δp_{HX})*

Measuring the pressure loss of a cooling heat exchanger is a method that is appropriate for the stability testing of PCS. An unstable PCS (e.g., damaged microcapsules) releases PCM material, which crystallizes at the heat exchanger surface when cooling below the crystallization temperature. PCM crystallizing at the heat exchanger surface increases the pressure drop over the heat exchanger. Therefore, this technique is a proper inline detection for a possible degradation of a PCS.

3.12. *Particle Size Analysis (PSA)*

PCS show a change in particle size distribution when they degrade. Particle size measurement can be done inline or with separate devices (e.g., laser diffractometer). Using separate devices, samples have to be taken while the material is in a degradation process (e.g., conveyed in a hydraulic loop).

3.13. *Viscosity Measurement (μ)*

Viscosity measurements can be used to detect the degradation of PCS. A change in particle size is usually connected to a change in viscosity. However, even if there is no change in particle size, a reduction of the PCM fraction can occur due to separation in hydraulic systems. Such a separation leads to lower viscosity.

4. Conclusions

The present survey is based on 18 experimental devices to investigate the long-term stability of solid–liquid PCM. The experiments are divided into thermal cycling stability tests, tests on PCM with stable supercooling, and tests on the stability of phase change slurries (PCS).

Comparing the technical specifications and the applied test conditions, a large variety can be observed among the mostly self-built test rigs. Most of the devices are being used to perform thermal cycling stability tests. Depending on the type of PCM to be investigated and the intended application, different testing conditions are chosen, such as temperature range of investigation, heating and cooling rates to perform thermal cycles, sample container materials, and the atmosphere that the PCM sample is in contact with. Sample sizes in the order of few milliliters allow for the quick screening of properties, whereas samples of several hundred milliliters and more allow evaluating the material's behavior without the enhanced supercooling typically observed in measurements on small samples. The chosen speed of PCM cycling depends on several boundary conditions, such as the sample size under investigation. Typically, the larger the sample, the slower the cycling. Relatively slow cycling of larger samples is often done in order to study the thermal cycling behavior of PCM under application conditions with few cycles per day to be carried out. So far, there is no consensus on how to accelerate stability tests of PCM reasonably.

While several test rigs are designed to investigate different types of PCM for different applications, tests on PCM with stable supercooling are dedicated—but not limited—to SAT. Applied temperatures (charging, discharging, and ambient) as well as the degree and the duration of supercooling should be in accordance to the conditions of the aimed application.

In addition to the solid–liquid phase transition, tests on PCS investigate the mechanical stability of capsules or emulsions during pumping the PCS.

Overall, the comparison of experimental devices can be an encouragement to discuss and develop uniform standards of PCM stability testing. The fact that many further devices are being used in other research groups strengthens this. A possible goal for future work in this context could be to propose a guideline for long-term stability investigations or to further develop existing testing regulations, e.g., the regulations from the "RAL Quality Association PCM". Investigating the stability of PCM under application conditions requires, first, knowledge about the application and, second, a method to translate the application requirements into experimental testing conditions. Thereby, the second step influences the technical design of testing devices and applied testing procedures. It is especially this "translation" that is nowadays being done in a highly non-standardized manner. Further work toward a standardization of PCM stability testing is recommended.

Author Contributions: Conceptualization, C.R., S.H. (Stefan Hiebler), R.B., L.F.C., G.Z., G.E., M.D., G.D., O.F., R.R., D.G., S.G., S.H. (Stephan Höhlein), A.K.-H. and L.Z.; Funding acquisition, C.R., S.H. (Stefan Hiebler), R.B., L.F.C., G.Z., G.E., M.D., G.D., O.F., R.R., D.G., S.G., S.H. (Stephan Höhlein), A.K.-H. and L.Z.; Investigation, C.R., S.H. (Stefan Hiebler), R.B., L.F.C., G.Z., G.E., M.D., G.D., O.F., R.R., D.G., A.C.K., S.G., S.H. (Stephan Höhlein), A.K.-H., N.B. and L.Z.; Methodology, C.R., S.H. (Stefan Hiebler), R.B., L.F.C., G.Z., G.E., M.D., G.D., O.F., R.R., D.G., S.G., S.H. (Stephan Höhlein), A.K.-H. and L.Z.; Writing—original draft, C.R.; Writing—review and editing, C.R., S.H. (Stefan Hiebler), R.B., L.F.C., G.Z., G.E., M.D., G.D., O.F., R.R., D.G., S.G., S.H. (Stephan Höhlein), A.K.-H. and L.Z. All authors have read and agreed to the published version of the manuscript.

Funding: This survey has been carried out within the framework of IEA ES Annex 33/SHC Task 58 "Material and Component Development for Compact Thermal Energy Storage", a joint working group of the "Energy Storage" (ES) and the "Solar Heating and Cooling" (SHC) Technology Collaboration Programmes of the International Energy Agency (IEA). The responsibility for the content of this publication is with the authors. The work of ZAE Bayern is part of the project properPCM and was supported by the German Federal Ministry of Economic Affairs and Energy under the project code 03ET1342A. The work of CIEMAT is part of two projects: ACES2030 (S2018/EM-4319) which is supported by Comunidad de Madrid and European Structural Funds through and SFERA III (GA No 823802) which is supported by European Union's Horizon H2020 Research and Innovation Programme. The work of DTU was supported by HM Heizkörper GmbH and the Danish Energy Agency through the EUDP program. The work of EHU is part of the project Sweet-TES (RTI2018-099557-B-C22), funded by the Spanish Ministry of Science, Innovation and Universities. The work of GREiA was partially funded by the Ministerio de Ciencia, Innovación y Universidades de España (RTI2018-093849-B-C31—MCIU/AEI/FEDER, UE) and by the Ministerio de Ciencia, Innovación y Universidades—Agencia Estatal de Investigación (AEI) (RED2018-102431-T). The authors would like to thank the Catalan Government for the quality accreditation given to their research group (GREiA 2017 SGR 1537). GREiA is certified agent TECNIO in the category of technology developers from the Government of Catalonia. This work is partially supported by ICREA under the ICREA Academia programme. The work of HSLU was partially funded by the Swiss Competence Centre for Energy Research on Heat and Electricity (SCCER HaE). The authors would like to thank the SCCER HaE as well as METTLER TOLEDO AG for the support given. The work of Fraunhofer ISE was part of the projects KOLAN (FKZ 03ESP357B) and KOKAP (FKZ 03ET1463A) which were administrated by the Projektträger Jülich (PtJ) and funded by the German Federal Ministry of Economic Affairs and Energy. The work of LAMTE was supported by Intel Corporation, along with the Natural Sciences and Engineering Research Council of Canada (NSERC) and the Canada Foundation for Innovation. The work of LGCgE is partly funded by the French National Research Agency (ANR) under the project "Hybrid storage and heat exchanger device with PCM—EUROPA". Noé Beaupere's PhD work was funded by the French Alternative Energies and Atomic Energy Commission (CEA) Grenoble. The work of LTTT is part of the project MALATrans and was funded by the German Federal Ministry for Economic Affairs and Energy (BMWi) under the project code 03ESP227B.

Conflicts of Interest: The authors declare no conflict of interest. The funders had no role in the design of the study; in the collection, analyses, or interpretation of data; in the writing of the manuscript, or in the decision to publish the results.

Nomenclature

DSC	Differential scanning calorimetry
FT-IR	Fourier transform infrared spectroscopy
HTF	Heat transfer fluid
PCM	Phase change material
PCS	Phase change slurry
PMMA	Poly (methyl methacrylate)
PSA	Particle size analysis
RLM	Reflected light microscope
SAT	Sodium acetate trihydrate
TES	Thermal energy storage
TGA	Thermogravimetric analysis
TCM	Thermochemical material
XRD	X-ray diffraction spectroscopy
h	Specific enthalpy/J·g^{-1}
m	Mass/kg
μ	Viscosity/Pa·s
T	Temperature/°C
t	Time/s

References

1. Mehling, H.; Cabeza, L.F. *Heat and Cold Storage with PCM*; Springer: Berlin, Germany, 2008.
2. Zalba, B.; Marın, J.M.; Cabeza, L.F.; Mehling, H. Review on thermal energy storage with phase change: Materials, heat transfer analysis and applications. *Appl. Therm. Eng.* **2003**, *23*, 251–283.
3. Cabeza, L.F.; Castell, A.; Barreneche, C.D.; De Gracia, A.; Fernández, A.I. Materials used as PCM in thermal energy storage in buildings: A review. *Renew. Sustain. Energy Rev.* **2011**, *15*, 1675–1695.
4. Zsembinszki, G.; Fernández, A.G.; Cabeza, L.F. Selection of the Appropriate Phase Change Material for Two Innovative Compact Energy Storage Systems in Residential Buildings. *Appl. Sci.* **2020**, *10*, 2116.
5. IEA ES Annex 33/SHC Task 58 "Material and Component Development for Compact Thermal Energy Storage". Available online: https://task58.iea-shc.org/ (accessed on 17 August 2020).
6. Ferrer, G.; Solé, A.; Barreneche, C.; Martorell, I.; Cabeza, L.F. Review on the methodology used in thermal stability characterization of phase change materials. *Renew. Sustain. Energy Rev.* **2015**, *50*, 665–685.
7. Rathgeber, C.; Grisval, A.; Schmit, H.; Hoock, P.; Hiebler, S. Concentration dependent melting enthalpy, crystallization velocity, and thermal cycling stability of pinacone hexahydrate. *Thermochim. Acta* **2018**, *670*, 142–147.
8. Bayón, R.; Rojas, E. Development of a new methodology for validating thermal storage media: Application to phase change materials. *Int. J. Energy Res.* **2019**, *43*, 6521–6541.
9. RAL Quality Association PCM. Quality and Testing Specifications for Phase Change Materials. 2018. Available online: http://www.pcm-ral.org/pcm/en/quality-testing-specifications-pcm/ (accessed on 17 August 2020).
10. Englmair, G.; Jiang, Y.; Dannemand, M.; Moser, C.; Schranzhofer, H.; Furbo, S.; Fan, J. Crystallization by local cooling of supercooled sodium acetate trihydrate composites for long-term heat storage. *Energy Build.* **2018**, *180*, 159–171.
11. Desgrosseilliers, L. Design and Evaluation of a Modular, Supercooling Phase Change Heat Storage Device for Indoor Heating. Ph.D. Thesis, Dalhousie University, Halifax, NS, Canada, 2016. Available online: https://dalspace.library.dal.ca/handle/10222/73309 (accessed on 17 August 2020).
12. Englmair, G.; Moser, C.; Furbo, S.; Dannemand, M.; Fan, J. Design and functionality of a segmented heat-storage prototype utilizing stable supercooling of sodium acetate trihydrate in a solar heating system. *Appl. Energy* **2018**, *221*, 522–534.

13. Englmair, G.; Kong, W.; Berg, J.B.; Furbo, S.; Fan, J. Demonstration of a solar combi-system utilizing stable supercooling of sodium acetate trihydrate for heat storage. *Appl. Therm. Eng.* **2020**, *166*, 114647.
14. Englmair, G.; Moser, C.; Schranzhofer, H.; Fan, J.; Furbo, S. A solar combi-system utilizing stable supercooling of sodium acetate trihydrate for heat storage: Numerical performance investigation. *Appl. Energy* **2019**, *242*, 1108–1120.
15. Duquesne, M.; Del Barrio, E.P.; Godin, A. Nucleation triggering of highly undercooled Xylitol using an air lift reactor for seasonal thermal energy storage. *Appl. Sci.* **2019**, *9*, 267.
16. Puupponen, S.; Mikkola, V.; Ala-Nissila, T.; Seppälä, A. Novel microstructured polyol-polystyrene composites for seasonal heat storage. *Appl. Energy* **2016**, *172*, 96–106.
17. Rathod, M.K.; Banerjee, J. Thermal stability of phase change materials used in latent heat energy storage systems: A review. *Renew. Sustain. Energy Rev.* **2013**, *18*, 246–258.
18. Bayón, R.; Biencinto, M.; Rojas, E.; Uranga, N. Study of hybrid dry cooling systems for STE plants based on latent storage. In Proceedings of the ISEC Conference, Graz, Austria, 3–5 October 2018.
19. Rodríguez-García, M.M.; Rojas, E.; Bayón, R. Test campaign and performance evaluation of a spiral latent storage module with Hitec® as PCM. In Proceedings of the SWC 2017/SHC2017 Conference, Abu Dhabi, UAE, 29 October–2 November 2017.
20. Rodríguez-García, M.M.; Bayón, R.; Rojas, E. Stability of D-mannitol upon melting/freezing cycles under controlled inert atmosphere. *Energy Procedia* **2016**, *91*, 218–225.
21. Biedenbach, M.; Klünder, F.; Gschwander, S. Investigations on the stability of metallic cans for PCM macro-encapsulation. In Proceedings of the International Institute of Refrigeration (IIR), Birmingham, AL, USA, 2–5 September 2018.
22. Kahwaji, S.; Johnson, M.B.; Kheirabadi, A.C.; Groulx, D.; White, M.A. Stable, low-cost phase change material for building applications: The eutectic mixture of decanoic acid and tetradecanoic acid. *Appl. Energy* **2016**, *168*, 457–464.
23. Kheirabadi, A.C.; Groulx, D. Design of an automated thermal cycler for long-term phase change material phase transition stability studies. In Proceedings of the COMSOL Conference 2014, Boston, MA, USA, 8–10 October 2014.
24. Kahwaji, S.; Johnson, M.B.; Kheirabadi, A.C.; Groulx, D.; White, M.A. Fatty acids and related phase change materials for reliable thermal energy storage at moderate temperatures. *Sol. Energy Mater. Sol. Cells* **2017**, *167*, 109–120.
25. Kahwaji, S.; Johnson, M.B.; Kheirabadi, A.C.; Groulx, D.; White, M.A. A comprehensive study of properties of paraffin phase change materials for solar thermal energy storage and thermal management applications. *Energy* **2018**, *162*, 1169–1182.
26. Beaupere, N. Pilotage de la Libération de Chaleur et Etude du Vieillissement de Matériaux à Changement de Phase. Ph.D. Thesis, Université d'Artois/CEA Grenoble, Grenoble, France, 2019.
27. Orak, F.; König-Haagen, A.; Brüggemann, D. Effect of Nucleators on Heat Storage Properties of Salt Hydrates. In Proceedings of the 2017 11th International Renewable Energy Storage Conference (IRES 2017), Düsseldorf, Germany, 14–16 March 2017.
28. Navarro, L.; Solé, A.; Martín, M.; Barreneche, C.; Olivieri, L.; Tenorio, J.A.; Cabeza, L.F. Benchmarking of useful phase change materials for a building application. *Energy Build.* **2019**, *182*, 45–50.
29. Oró, E.; Gil, A.; Miró, L.; Peiró, G.; Álvarez, S.; Cabeza, L.F. Thermal energy storage implementation using phase change materials for solar cooling and refrigeration applications. *Energy Procedia* **2012**, *30*, 947–956.
30. Gil, A.; Oró, E.; Castell, A.; Cabeza, L.F. Experimental analysis of the effectiveness of a high temperature thermal storage tank for solar cooling applications. *Appl. Therm. Eng.* **2013**, *54*, 521–527.
31. Gil, A.; Oró, E.; Miró, L.; Peiró, G.; Ruiz, A.; Salmerón, J.M.; Cabeza, L.F. Experimental analysis of hydroquinone used as phase change material (PCM) to be applied in solar cooling refrigeration. *Int. J. Refrig.* **2014**, *39*, 95–103. [CrossRef]
32. Peiró, G.; Gasia, J.; Miró, L.; Cabeza, L.F. Experimental evaluation at pilot plant scale of multiple PCMs (cascaded) vs. single PCM configuration for thermal energy storage. *Renew. Energy* **2015**, *83*, 729–736. [CrossRef]

33. Prieto, C.; Miró, L.; Peiró, G.; Oró, E.; Gil, A.; Cabeza, L.F. Temperature distribution and heat losses in molten salts tanks for CSP plants. *Sol. Energy* **2016**, *135*, 518–526. [CrossRef]
34. Peiró, G.; Gasia, J.; Miró, L.; Prieto, C.; Cabeza, L.F. Experimental analysis of charging and discharging processes, with parallel and counter flow arrangements, in a molten salts high temperature pilot plant scale setup. *Appl. Energy* **2016**, *178*, 394–403. [CrossRef]
35. Gasia, J.; de Gracia, A.; Peiró, G.; Arena, S.; Cau, G.; Cabeza, L.F. Use of partial load operating conditions for latent thermal energy storage management. *Appl. Energy* **2018**, *216*, 234–242. [CrossRef]
36. Peiró, G.; Prieto, C.; Gasia, J.; Jové, A.; Miró, L.; Cabeza, L.F. Two-tank molten salts thermal energy storage system for solar power plants at pilot plant scale: Lessons learnt and recommendations for its design, start-up and operation. *Renew. Energy* **2018**, *121*, 236–248. [CrossRef]
37. Gil, A.; Peiró, G.; Oró, E.; Cabeza, L.F. Experimental analysis of the effective thermal conductivity enhancement of PCM using finned tubes in high temperature bulk tanks. *Appl. Therm. Eng.* **2018**, *142*, 736–744. [CrossRef]
38. Gasia, J.; de Gracia, A.; Zsembinszki, G.; Cabeza, L.F. Influence of the storage period between charge and discharge in a latent heat thermal energy storage system working under partial load operating conditions. *Appl. Energy* **2019**, *235*, 1389–1399. [CrossRef]
39. Mselle, B.D.; Verez, D.; Zsembinszki, G.; Borri, E.; Cabeza, L.F. Performance Study of Direct Integration of Phase Change Material into an Innovative Evaporator of a Simple Vapour Compression System. *Appl. Sci.* **2020**, *10*, 4649. [CrossRef]
40. Dannemand, M.; Dragsted, J.; Fan, J.; Johansen, J.B.; Kong, W.; Furbo, S. Experimental investigations on prototype heat storage units utilizing stable supercooling of sodium acetate trihydrate mixtures. *Appl. Energy* **2016**, *169*, 72–80. [CrossRef]
41. Dannemand, M.; Johansen, J.B.; Kong, W.; Furbo, S. Experimental investigations on cylindrical latent heat storage units with sodium acetate trihydrate composites utilizing supercooling. *Appl. Energy* **2016**, *177*, 591–601. [CrossRef]
42. Englmair, G.; Furbo, S.; Dannemand, M.; Fan, J. Experimental investigation of a tank-in-tank heat storage unit utilizing stable supercooling of sodium acetate trihydrate. *Appl. Therm. Eng.* **2020**, *167*, 114709. [CrossRef]
43. Kong, W.; Dannemand, M.; Johansen, J.B.; Fan, J.; Dragsted, J.; Englmair, G.; Furbo, S. Experimental investigations on heat content of supercooled sodium acetate trihydrate by a simple heat loss method. *Sol. Energy* **2016**, *139*, 249–257. [CrossRef]
44. Dannemand, M.; Furbo, S. Supercooling stability of sodium acetate trihydrate composites in multiple heat storage units. *Refrig. Sci. Technol.* **2018**, 227–231. [CrossRef]
45. Niedermaier, S.; Biedenbach, M.; Gschwander, S. Characterisation and enhancement of phase change slurries. *Energy Procedia* **2016**, *99*, 64–71. [CrossRef]
46. Peiró, G. High Temperature Thermal Energy Storage for Solar Cooling and Solar Thermal Power Plants Applications. Ph.D. Thesis, University of Lleida, Lleida, Spain, 2017. Available online: http://hdl.handle.net/10803/462071 (accessed on 17 August 2020).
47. Deng, J.; Furbo, S.; Kong, W.; Fan, J. Thermal performance assessment and improvement of a solar domestic hot water tank with PCM in the mantle. *Energy Build.* **2018**, *172*, 10–21. [CrossRef]
48. Gschwander, S.; Haussmann, T.; Hagelstein, G.; Barreneche, C.; Ferrer, G.; Cabeza, L.; Diarce, G.; Hohenauer, W.; Lager, D.; Rathgeber, C.; et al. Standardization of pcm characterization via DSC. *Refrig. Sci. Technol.* **2016**, 70–75. [CrossRef]
49. Rathgeber, C.; Schmit, H.; Miró, L.; Cabeza, L.F.; Gutierrez, A.; Ushak, S.N.; Hiebler, S. Enthalpy-temperature plots to compare calorimetric measurements of phase change materials at different sample scales. *J. Storage Mater.* **2018**, *15*, 32–38. [CrossRef]
50. Rathgeber, C.; Miró, L.; Cabeza, L.F.; Hiebler, S. Measurement of enthalpy curves of phase change materials via DSC and T-History: When are both methods needed to estimate the behaviour of the bulk material in applications? *Thermochim. Acta* **2014**, *596*, 79–88. [CrossRef]
51. Bayón, R.; Rojas, R. Feasibility study of D-mannitol as phase change material for thermal storage. *AIMS Energy* **2017**, *5*, 404–424. [CrossRef]

52. Doyle, C.D. Estimating isothermal life from thermogravimetric data. *J. Appl. Polym. Sci.* **1962**, *6*, 639–642. [CrossRef]
53. Solé, A.; Neumann, H.; Niedermaier, S.; Martorell, I.; Schossig, P.; Cabeza, L.F. Stability of sugar alcohols as PCM for thermal energy storage. *Sol. Energy Mater. Sol. Cells* **2014**, *126*, 125–134. [CrossRef]

Publisher's Note: MDPI stays neutral with regard to jurisdictional claims in published maps and institutional affiliations.

Article

Experimental Analysis on the Thermal Management of Lithium-Ion Batteries Based on Phase Change Materials

Mingyi Chen [1,*], Siyu Zhang [1], Guoyang Wang [1], Jingwen Weng [2], Dongxu Ouyang [2], Xiangyang Wu [1], Luyao Zhao [1] and Jian Wang [2]

1 School of the Environment and Safety Engineering, Jiangsu University, Zhenjiang 212013, China; 2211909027@stmail.ujs.edu.cn (S.Z.); 2211909026@stmail.ujs.edu.cn (G.W.); wuxy@ujs.edu.cn (X.W.); zhaoly@ujs.edu.cn (L.Z.)

2 State Key Laboratory of Fire Science, University of Science and Technology of China, Hefei 230026, China; wengjw@mail.ustc.edu.cn (J.W.); ouyang11@mail.ustc.edu.cn (D.O.); wangj@ustc.edu.cn (J.W.)

* Correspondence: chenmy@ujs.edu.cn

Received: 10 September 2020; Accepted: 19 October 2020; Published: 21 October 2020

Abstract: Temperature is an important factor affecting the working efficiency and service life of lithium-ion battery (LIB). This study carried out the experiments on the thermal performances of Sanyo ternary and Sony $LiFePO_4$ batteries under different working conditions including extreme conditions, natural convection cooling and phase change material (PCM) cooling. The results showed that PCM could absorb some heat during the charging and discharging process, effectively reduce the temperature and keep the capacity stable. The average highest temperature of Sanyo LIB under PCM cooling was about 54.4 °C and decreased about 12.3 °C compared with natural convection in the 2 C charging and discharging cycles. It was found that the addition of heat dissipation fins could reduce the surface temperature, but the effect was not obvious. In addition, the charge and discharge cycles of the two kinds of LIBs were compared at the discharge rates of 1 C and 2 C. Compared with natural convection cooling, the highest temperature of Sanyo LIB with PCM cooling decreased about 4.7 °C and 12.8 °C for 1 C and 2 C discharging respectively, and the temperature of Sony LIB highest decreased about 1.1 °C and 2 °C.

Keywords: lithium-ion battery; thermal management; phase change material; temperature; heat dissipation fins; capacity

---

1. Introduction

The ways to deal with the energy crisis and environmental pollution and develop new energy of safety and cleanness have become the focus of the attention of all countries in the world. It also becomes the core issue to solve the sustainable development of humankind. LIBs, which have the advantages of high specific energy, relatively high operating voltage, long cycle life, low self-discharge rate, small and convenient, are considered to be one of the most promising energy storage devices [1–3]. However, if the operating temperature of LIB is high and the heat dissipation is not timely, it will lead to thermal failure, shorten the service life of LIB and cause safety problems, such as fire and explosion [4,5]. Therefore, the thermal safety of LIB needs to be focused on.

The temperature difference between battery internal and ambient, as well as the temperature difference between the cells inside the battery pack will have a negative impact on the performance, life and safety of the battery [6]. Therefore, it is necessary to conduct a reasonable heat management system to make the LIB working in the normal operating temperature range [7–10]. At present, there are three kinds of mainstream thermal management methods for LIBs including air cooling, liquid cooling and PCM cooling [11]. Air cooling is divided into natural air cooling and forced air cooling. Pesaran et al. studied the air cooling performance of the battery thermal management system (BTMS). The results showed that the air cooling could effectively reduce the battery temperature and keep the battery temperature consistent in a low rate discharging [12]. Giuliano et al. studied the thermal management system using air cooling of metal foam-based heat exchanger plate. The system showed that the LIB temperature decreased with the increase of airflow. However, the thermal conductivity of air is low and the heat dissipation is weak [13]. Compared with the air cooling method, the liquid cooling method has higher thermal conductivity, which leads to higher cooling performance and is more suitable for cooling large battery packs. Liquid cooling is divided into direct and indirect liquid cooling. Chen et al. compared different liquid cooling systems of LIB, and the results showed that the indirect liquid cooling system had the lowest temperature rise and was more practical than direct liquid cooling [14]. Zhang et al. applied the S-type guide plate to the integrated module of liquid cooling heat dissipation system, and found that the device could avoid heat concentration, improve heat transfer performance and keep the temperature of LIB uniform [15]. However, the traditional battery thermal management method that uses cheap air and water as the cooling medium requires additional energy to drive the cooling medium circulation. In battery thermal management applications, the battery's safe operating temperature is often sacrificed at the expense of battery capacity and power in exchange for longer operating life, and these traditional thermal management methods are complex and occupy a large space [16]. In recent years, researchers have found that PCMs exhibit excellent performance in thermal management of LIB [17,18] since they have large storage density apart of many other advantages [19–21]. Duan et al. conducted the numerical and experimental studies on PCM heat transfer, and found that they have great potential in thermal management due to potential energy storage and controllable temperature stability [22]. PCM battery management system which performed better than traditional thermal management system was first proposed and patented by Al-Hallaj and Selman [23,24]. Mills and Al-Hallaj used the entropy coefficient method to simulate the battery pack, and the results showed that using a PCM could significantly improve the system performance and keep the operating temperature below 55 °C, even at high discharge rate [25]. Javani et al. studied the effect of a PCM on LIB, and found that it could make the temperature distribution uniform and keep the battery in a safe temperature range [26]. Weng et al. systematically investigated the cooling behavior and the influence of several detailed factors on the performance of a PCM. The experimental results showed that a PCM module with a thickness of ~10 mm presented the optimal cooling performance [27]. Javani et al. investigated the heat transfer with a PCM in passive thermal management of electric and hybrid electric vehicles. The results showed that the temperature distribution became about 10% more uniform when a PCM was applied in a 3 mm thickness around the cell. The PCM with 12 mm thickness decreased the maximum temperature about 3.04 °C [28]. However, it is found that PCM is easy to get heat saturation when absorbing a large amount of heat in the practical application. BTMS based on pure phase-varying materials cannot work for a long time in high-power batteries and effectively control the battery temperature. Commonly used PCM is paraffin wax whose thermal conductivity is very low. The thermal conductivity of PCM is improved by adding high thermal conductivity materials such as expanded graphite, carbon fiber, graphene, aluminum foam, copper foam and so on [29–32].

On the other hand, many researchers combined other cooling methods with PCM-based BTMS in order to improve the cycling stability and thermal management ability [33,34]. The complex PCM based BTMS can effectively reduce temperature rise and temperature difference, and maintain battery performance, particularly in extreme environments [35,36]. Sun et al. proposed a PCM combined with heat-dissipating fins to enhance the heat transfer. The results showed that the performance of PCM-fin system was better than that of pure PCM system [37]. Azizi et al. used a PCM and composite materials of aluminum mesh plates to carry on the thermal management to the $LiFePO_4$ battery pack under the high temperature environment. The results showed that the usage of PCM and aluminum wire mesh could significantly reduce the surface temperature of the batteries and improve the performance of the battery pack [38]. Although the use of PCM either alone or combined with heat dissipation fins had been extensively studied, there is no optimal solution for the best cooling effect, quantity, and price of PCMs used in LIBs at present. The combination of a PCM and heat dissipation fin will inevitably increase the weight of the BTMS device. As far as we know, a PCM with high thermal conductivity in the battery charging and discharging cycles and the optimization of the coexistence system of PCM and heat dissipation fin also needs further research. In this paper, we first tested the cooling effect of a PCM with higher thermal conductivity whose phase change process is a solid-solid phase change, and then designed a new type of heat dissipation fins. The main purpose of this paper is to study the cooling effect of a PCM with high thermal conductivity on the battery and the effect of a PCM combined with a new type of heat dissipation fins with a relatively small volume. The authors conducted the experiments to investigate the temperature and capacity changing of two kinds of batteries under different conditions.

2. Experiment

2.1. Experimental Description

Experiments were carried out on two different brands of LIB (Sanyo ternary and Sony $LiFePO_4$ battery). The detailed parameters of the batteries were shown in Table 1. The experimental conditions of the battery in extreme conditions (inside the closed aluminum tube), natural convection, pure PCM, and PCM combined with the heat dissipation fins were set up. The experiments were carried out in the carton box (length × width × height, 35 × 20 × 20 cm) for the thermal isolation from environmental influences, and each experimental condition is shown in Figure 1. The aluminum tube size with an inner diameter of 36 mm, the height of 65 mm, and thickness of 6 mm (provided by Yuezhong Metal Material Co., Ltd., Dongguan, China) was used, and its thermal conductivity is 237.2 W/m·K. Figure 1d shows that the fin is composed of two circular fins and three vertical fins. Its material is tin sheet and the thermal conductivity is 150 W/m·K. The design of the fin structure is followed by: (1) The ring is set at the bottom of the battery and in the middle of the battery respectively as the temperature of the middle and low part of the battery is higher. (2) The vertical fin protruding is designed to disperse the generated heat in time and reduce the temperature effect. (3) According to previous studies, the best results are obtained when the distance between the radiator and the battery is 0.2 times the diameter of the battery [39]. So, the circular heat dissipation fin size (outer diameter × inner diameter × height) is $\varphi 26 \times \varphi 22 \times 4$ mm, vertical heat dissipation fin size (length × width × height) is 4 × 2 × 80 mm.

The high thermal conductivity phase change composites with phase change temperature of 52 °C (provided by Zhongjia New Material Technology Co., Ltd., Guangzhou, China) were selected due to the higher thermal conductivity than the pure paraffin. In this experiment, the phase change material is wrapped around the battery with a thickness of 9 mm and a weight of about 9.9 g. The main components are phase change wax and expanded graphite, which have the following advantages: solid-solid phase change, high phase change latent heat, insulation, good cycle stability, non-corrosive, and non-toxic. Detailed thermal physical properties are shown in Table 2.

Table 1. Detailed battery parameters.

Battery Type	Sanyo Lithium-Ion Battery	Sony Lithium-Ion Battery
Capacity	2600 mAh	1100 mAh
Voltage (V)	3.7 V (Discharge cut-off voltage 3.0 V, The maximum charging voltage is 4.2 V	3.2 V (Discharge cut-off voltage 2.0 V, The maximum charging voltage is 3.6 V)
Sizes	Φ18 × 65 mm	Φ18 × 65 mm
Life (seconds)	1000	1000
Anode material	Ternary materials	$LiFePO_4$

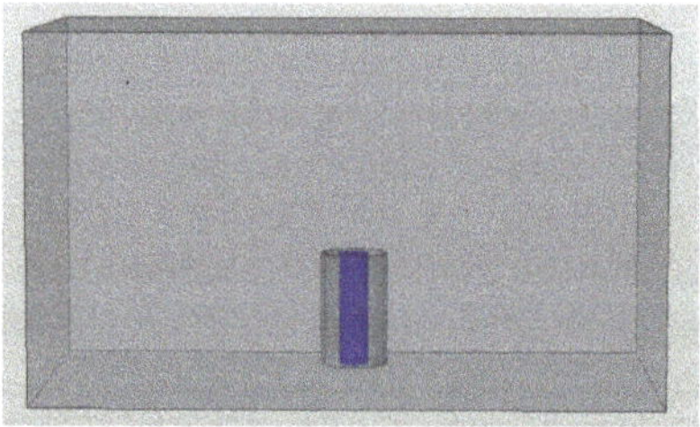

(**a**) Extreme conditions (charge/discharge inside the closed aluminum tube)

(**b**) Natural convection cooling

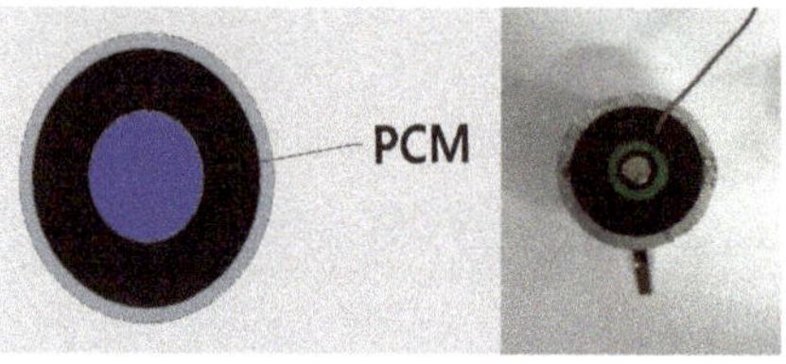

(**c**) Phase change material cooling

(**d**) Heat dissipation fins, pcm-fin cooling

Figure 1. Experimental conditions

Table 2. Thermal physical properties of PCM.

Thermal Physical Characteristic	Parameters
Phase change temperature (°C)	52~55
Thermal conductivity (W/m·K)	3
Latent heat of phase change (J/g)	170
Specific heat capacity (J/kg·°C)	1.8
Density (g/cm^3)	0.2
Materials	Phase change wax, Expanded graphite

2.2. Experimental Facility

Figure 2 shows the schematic diagram of the experiments. The LIB charging and discharging instrument (provided by Neware Electronics Co., Ltd., Shenzhen, China) is provided for the LIB normal charging and discharging cycles, and the accuracy and stability of the whole range can reach 0.1%. The LIB temperature was measured by T-type thermocouples (provided by OMEGA), and recorded by the temperature data acquisition unit (provided by National Instruments) as shown in Figure 2. According to the factory temperature check, the thermocouple temperature measurement accuracy is controlled within 0.5 °C. The thermocouple is patch type, which is attached to the middle of the battery surface. The PCM is wrapped around the battery, and the thermocouple is embedded in the PCM. The thermocouple was calibrated before the experiments. The temperature of the LIB by thermocouple was measured only to analyze the cooling effect of PCM on the LIB, not to study the internal reaction of the LIB in this study. Hence, the T-type thermocouple is attached to the center position of the batteries surface to measure the optimum temperature.

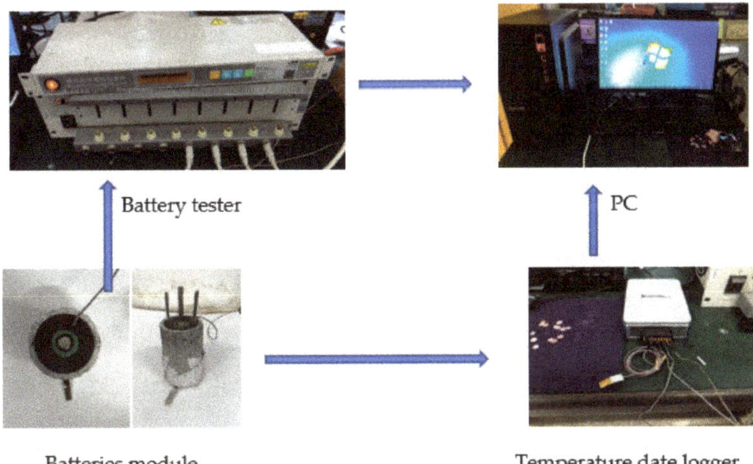

Figure 2. Schematic diagram of the experiment.

Before the experiment, the LIBs were discharged to the cut-off voltage, charged to maximum voltage, and then placed 24 h to keep the LIB stable. The temperature in the carton is 25 °C, and its environmental temperature error is controlled within the normal range of 0.5 °C. The charge and discharge process of the two batteries are shown in Table 3. The cycle of Sanyo LIB can be divided into four stages: (1) discharge stage, with 2 C constant current discharge to 2.5 V. (2) shelving stage for 5 min. (3) charging stage with 1 C constant current and voltage to 4.2 V. (4) shelving stage with 5 min. The Sony LIB is tested in the same charge-discharge cycle as the Sanyo LIB. The only differences are the cutoff discharging voltage of 2.0 V and cutoff charging voltage of 3.6 V because of its relatively small capacity and voltage. Besides, two kinds of cells are placed in the PCM to discharge at a discharge rate of 1 C and 2 C.

Table 3. Design of charging and discharging experimental conditions.

Processing	Condition	Current (A)		Voltage (V)	
		Sanyo Battery	Sony Battery	Sanyo Battery	Sony Battery
Discharging	Constant current	2.6 (1C), 5.2 (2C)	1 (1C),2 (2C)	4.2	3.6
shelving	5 min				
Charging	Constant voltage and current	2.6 (1C)	1 (1C)	2.5	2.0
shelving	5 min				

3. Results and Discussions

3.1. Temperature Change during LIB Cycle

When the LIB is charging and discharging, the reaction occurs inside the LIB, which produces heat and shows the increase of battery temperature. Figure 3 shows the temperature change during charging and discharging of Sanyo LIB under four different working conditions: extreme conditions, natural convection cooling, PCM cooling, and PCM combined with heat dissipation fins cooling at the ambient temperature of 25 °C. Figure 3a shows the temperature curve of LIB in two cycles. That can be seen from the figure, the temperature of LIB in the enclosed space rises from 30 °C to about 77.3 °C during discharging, which obviously exceeds the normal operating temperature of the battery. The high temperature has certain bad influence on the LIB. The long-term working under the extreme condition will cause the LIB internal electrolyte decomposition, the positive and negative electrode reaction, and so on, which may cause the LIB thermal runaway and produce the safety risks. The highest temperature of the LIB is about 54.5 °C under PCM cooling, 66.2 °C under natural convection, 77.3 °C under extreme condition. Compared to the air cooling, the temperature of LIB using PCM cooling has a much smaller increase rate, and the highest temperature drops by 11.7 °C. Therefore, the results show that PCM can absorb some heat during the LIB discharging process, effectively reducing the LIB temperature. For the analysis of the cooling condition of the PCM with heat dissipation fins on the LIB, the heat dissipation fins have been adopted. The volume ratio of PCM and heat dissipation fins in the device is about 96%:4%. According to the diagram, the maximum temperature difference between the LIB in this working condition and PCM cooling condition is about 0.2 °C. The combination of heat dissipation fins can slightly reduce the temperature of LIB. The reason may be the fin structure cannot dissipate the latent heat in time due to its small volume ratio. In the research on the combination of PCM with heat dissipating fins, while giving full play to the maximum effect of heat dissipating fins, its cost and influence on the whole LIB structure should also be considered. Javani et al. [28] investigated heat transfer with PCMs in passive thermal management of electric and hybrid electric vehicles. Their results showed that the temperature distribution became about 10% more uniform when the PCM was applied in a 3 mm thickness around the LIB. The PCM with 12 mm thickness decreased the maximum temperature about 6.6 °C. In this research, the PCM not only has better cooling performance, but also makes the device having the advantages of small size and lightweight. In addition, the PCM in this study has a solid-solid phase change process, which avoids the liquid leakage. The device structure in this study is simpler and more practical, which may has a potential for commercialization.

Figure 3b shows the temperature change curve of LIB during six cycles. The temperature of LIB tends to be stable under various working conditions. However, the average of the maximum temperatures for the six cycles of LIB reaches about 77.8 °C in the enclosed space, which is in an unsafe state. It may cause great damage to the internal structure of the LIB and seriously affect the cycle service life of LIB. The PCM can have a good cooling effect on the temperature and make the average of the maximum temperatures for the six cycles of LIB keep at about 54.4 °C. The average of the maximum temperatures for the six cycles of LIB under the experimental condition of the combination of PCM and heat dissipating fins is about 54.3 °C, which is not significantly different compared to PCM cooling condition. The result shows that the effect of heat dissipation fins is not significant and the pure PCM cooling is the best choice due to the cost and complexity of the device structure. On the other hand, the LIB temperature rose at a slower rate and the temperature curve shifted slightly to the right as the cycle progressed. The heat generated inside LIB consisted of five parts: electrochemical reaction heat, ohmic internal resistance heat, polarization heat, electrolytic decomposition heat, and SEI decomposition heat. The electrolytic and SEI decomposition heat were very small and could be ignored when LIB was in the normal operating temperature range. The LIB would produce heat during the charging and discharging cycles. For discharging condition, the total chemical reaction was an exothermic reaction and the temperature rose theoretically, and for charging condition, the total chemical reaction was an endothermic reaction and the temperature drops theoretically. However, in the case of natural convection in Figure 3b, the temperature remains unchanged or even rises within a short period of time (region A) during the battery charging process. In the LIB charge process, although it is an endothermic process, the LIB generated heat such as ohmic internal resistance heat, polarization internal resistance heat, and electrochemical reaction heat more than itself heat dissipation, so that the temperature remains stable or even slightly rises.

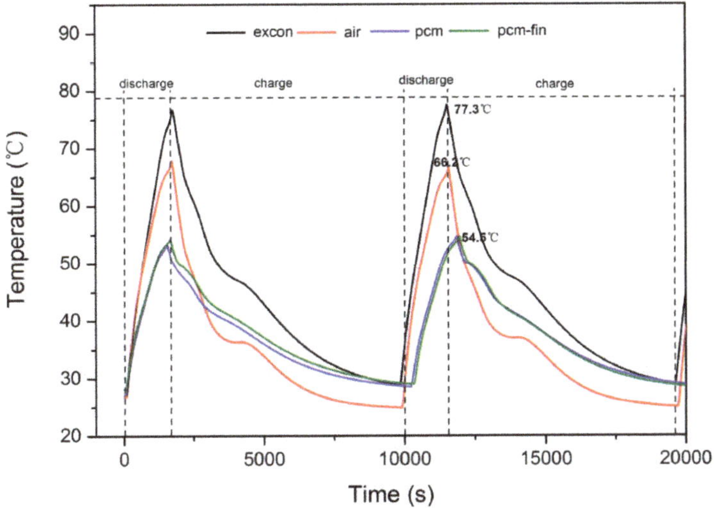

(**a**) Change of charging and discharging temperature of Sanyo LIB

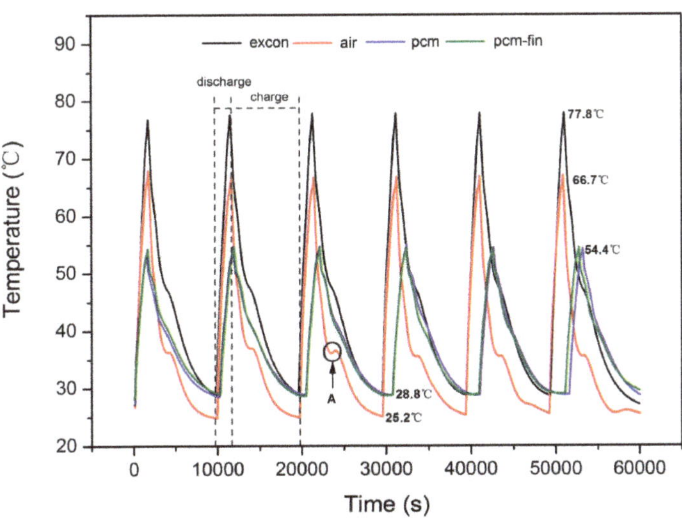

(**b**) The temperature changing of Sanyo LIB cycle

Figure 3. The temperature changing of LIBs during charging and discharging.

Figure 4 shows the temperature changes of $LiFePO_4$ battery in charging and discharging cycles under four different working conditions. It can be found that the highest temperature of the LIB under extreme conditions is about 40.6 °C and the lowest temperature is about 35.1 °C when the LIB is shelved after charging. The temperature of $LiFePO_4$ LIB will not be too high compared with Sanyo LIB when it works for a short time under extreme conditions due to its smaller LIB capacity, and lower internal polarization and ohmic resistance heat. The maximum temperature of Sony LIB is about 35.1 °C under PCM cooling, 37.5 °C under natural convection, 40.6 °C under extreme condition. It is found that the LIB temperature under PCM cooling decreases by 2.4 °C compared with the natural convection. In the case of the combination of PCM and heat dissipation fins, the temperature decreases 0.4 °C than PCM cooling due to the temperature of battery itself is low. In addition, the Sony LIB temperature in extreme conditions cannot be lowered to the level of other conditions during charging and shelving due to the heat is not dispersed in time. It can accelerate the degradation of the LIB and influence the LIB performance in extreme condition for a long time, which further prove the necessity of BTMS.

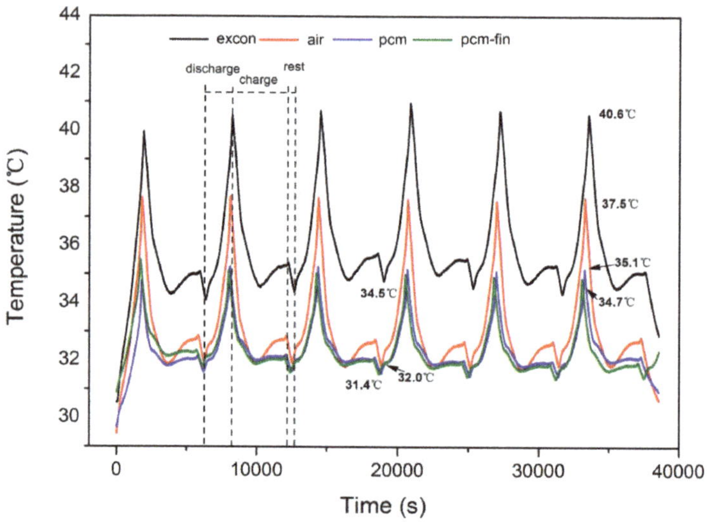

Figure 4. The temperature changes of $LiFePO_4$ battery during charging and discharging.

3.2. *The Influence of Temperature on Battery Performance*

Temperature is an important factor affecting battery performance parameters. Too high or low temperature will affect LIB capacity, charging and discharging efficiency, safety performance and result in battery performance degradation. Figure 5a shows the current change of the LIB in the charging and discharging cycle, and Figure 5b shows the voltage change. For the convenience of statistics, the x-coordinate in the figure represents the time step, and each step is 10 s. That can be seen from the figures, the current and voltage curves of the LIB have the same trend in each working condition. The current and voltage can reach the set rating in each charging and discharging cycle. However, as the cycles progresses, compared with the LIB under extreme conditions, the LIB curves in other conditions shift to the right, which is very similar to Figure 3b. The reason is related to the battery charging and discharging time. Under constant current conditions, the total flux of lithium ions should be roughly similar (both as diffusion in the electrolyte, and total flux into and out of the active material particles), and environmental conditions have little effect on it. However, at lower temperatures, a higher concentration gradient is required to overcome the slower diffusivity to meet the required flux. The higher concentration gradient leads to higher over potential and uneven utilization of electrodes. In turn, this leads to faster reaching of the cut-off voltage when working at lower temperatures. It can be clearly seen in Figure 5a that the LIBs in PCM and PCM-Fin cooling experiments seem to be faster reaching the cutoff voltage than the extreme condition and air cooling experiments in the first discharge cycle when all batteries start at the same state of charge (SOC). On the contrary, a higher temperature of LIB means that a constant current can be provided for a longer period of time. That can be seen in Figure 5b, the voltage drop of LIBs at higher temperature is smaller. In addition, according to Figure 3, it is found that in the constant current charging stage of the battery, the temperature of the battery decreased rapidly in the extreme condition and air cooling experiments, and the temperature is lower than the other cases in a period of time due to the latent heat effect of PCM. It seems that the LIBs under PCM and PCM-Fin cooling conditions can be charged for a longer time in the later cycles than the other cases.

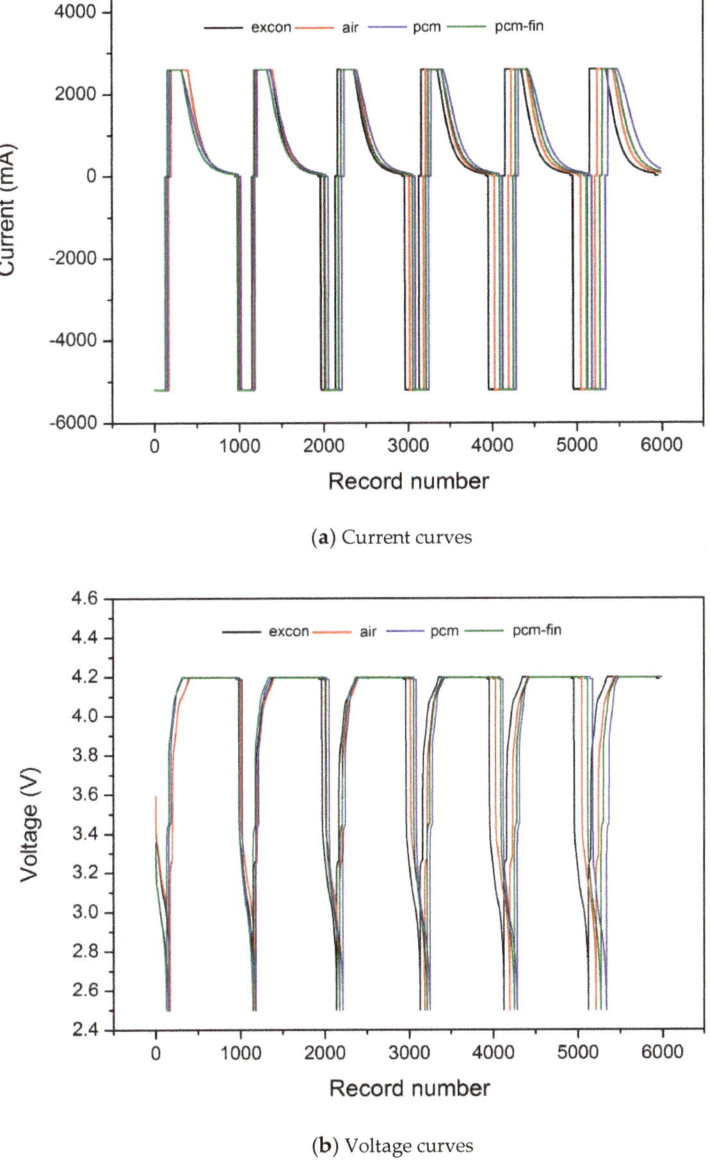

(**a**) Current curves

(**b**) Voltage curves

Figure 5. The current and voltage curves of Sanyo LIB.

Figure 6 shows the current and voltage changes in the $LiFePO_4$ battery during charging and discharging cycles. The current and voltage changes of $LiFePO_4$ LIB have the same trend under four different working conditions. Since the operating temperature of $LiFePO_4$ LIB under four working conditions is around 30–40 °C, the LIB is in a normal working state and its current and voltage will not change much.

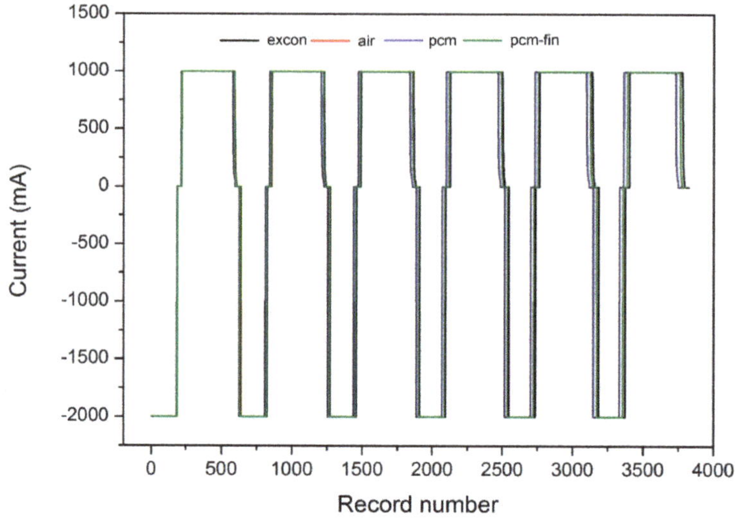

(**a**) Current curves

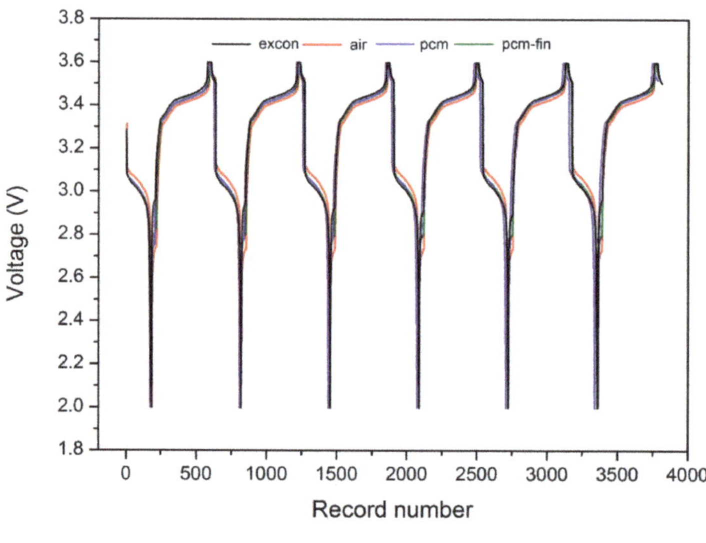

(**b**) Voltage curves

Figure 6. The current and voltage curves of Sony LIB.

Table 4 shows the capacity changes of Sanyo LIB charging and discharging cycles under different working conditions. It is found from the table that the LIB capacity has some changes in each working cycle. The average charging capacity of six cycles is 2414.6 mAh under the extreme conditions, 2399.4 mAh under natural convection, 2397.7 mAh under PCM cooling, and 2403.6 mAh under PCM combined with heat dissipation fins working condition. The average discharging capacity of six cycles is 2412.7 mAh under the extreme conditions, 2398.3 mAh under natural convection, 2398.4 mAh under PCM cooling, and 2383.3 mAh under PCM combined with heat dissipation fins working condition. Table 5 shows the capacity changes of $LiFePO_4$ battery charging and discharging cycles under different working conditions. The average charging capacity of six cycles is 1022.6, 1034.9, 1031.6, and 1034.5 mAh, and the average discharging capacity is 1026.8, 1035.9, 1030.4, and 1032.1 mAh, under the extreme conditions, natural convection, PCM cooling, and PCMcombined with heat dissipation fins working condition, respectively. The LIB capacity changes are related to the temperature. The battery capacity does not change significantly, which indicated that the LIB capacity is not significantly affected by short cycles even at high temperatures. Compared with the natural convection condition, the LIB discharging capacity under PCM cooling and PCM combined with heat dissipation fins decreased slightly. In the case of higher temperature, the battery reaches the cut-off voltage more slowly during the charging and discharging process, and the constant current charging and discharging time is longer, so the charging and discharging efficiency is higher, and the maximum capacity value is larger. However, the high temperature has a great influence on the capacity degradation rate of LIB [40]. The higher temperature leads to a quicker rate of various side reactions in the battery and faster the capacity degradation. Due to the low number of battery cycles in the experiments, the internal damage or possible side reactions that might have occurred at high temperatures had not yet become apparent. In the current application of LIB with the high efficiency and capacity, reducing the impact of itself side reaction and temperature are particularly critical, which is also a hurdle for the current LIB technology breakthrough.

Table 4. Capacity curve of Sanyo LIB in charge and discharge cycles.

Sanyo Battery Cycles	Excon		Air		PCM		PCM-fin	
	Charging Capacity (mAh)	Discharging Capacity (mAh)	Charging Capacity (mAh)	Charging Capacity (mAh)	Charging Capacity (mAh)	Charging Capacity (mAh)	Charging Capacity (mAh)	Charging Capacity (mAh)
1	2416.836	2413.664	2397.757	2396.380	2396.886	2385.235	2399.645	2387.914
2	2415.543	2412.135	2398.268	2398.719	2401.686	2483.068	2404.635	2379.632
3	2414.601	2413.030	2401.894	2400.978	2402.126	2379.272	2405.866	2384.824
4	2413.782	2412.072	2395.317	2399.563	2396.722	2382.227	2405.228	2386.124
5	2414.819	2413.401	2401.893	2396.380	2395.279	2384.336	2403.516	2378.967
6	2412.173	2411.815	2400.974	2397.655	2393.280	2376.504	2402.460	2382.489

Table 5. Capacity curve of $LiFePO_4$ LIB in charge and discharge cycles.

Sony Battery Cycles	Excon		Air		PCM		PCM-fin	
	Charging Capacity (mAh)	Discharging Capacity (mAh)	Charging Capacity (mAh)	Charging Capacity (mAh)	Charging Capacity (mAh)	Charging Capacity (mAh)	Charging Capacity (mAh)	Charging Capacity (mAh)
1	1031.890	1032.637	1035.779	1033.977	1032.103	1031.853	1035.741	1024.440
2	1008.673	1018.444	1033.233	1035.806	1032.359	1028.684	1031.933	1033.147
3	1018.732	1023.337	1034.633	1036.101	1030.632	1029.485	1034.568	1033.514
4	1022.702	1026.527	1035.109	1036.212	1031.216	1030.508	1035.078	1033.570
5	1025.826	1028.784	1035.155	1036.361	1031.566	1030.718	1034.831	1033.665
6	1028.002	1030.793	1035.320	1036.645	1031.840	1031.291	1035.110	1034.102

3.3. The PCM Influence on LIB Temperature under Different Discharging Ratios

Figure 7 shows the temperature curves of Sanyo and Sony LIB under the conditions of natural convection and PCM cooling. It is found that the PCM has a good effect to decrease the LIB temperature no matter it is discharged at a rate of 1 C or 2 C. Compared with natural convection cooling, the highest temperature of Sanyo LIB decreases about 4.7 °C and 12.8 °C for 1 C and 2 C discharging, respectively, and the highest temperature of Sony LIB decreases about 1.1 °C and 2 °C. The temperature reduction effect of PCM on high rate discharge LIB is more obvious. In the charging and discharging cycles, Sanyo LIB generates more heat due to its large capacity, and PCM absorbs more latent heat. The PCM has the most obvious effect on the Sanyo battery in a 2 C discharging rate among all experiments.

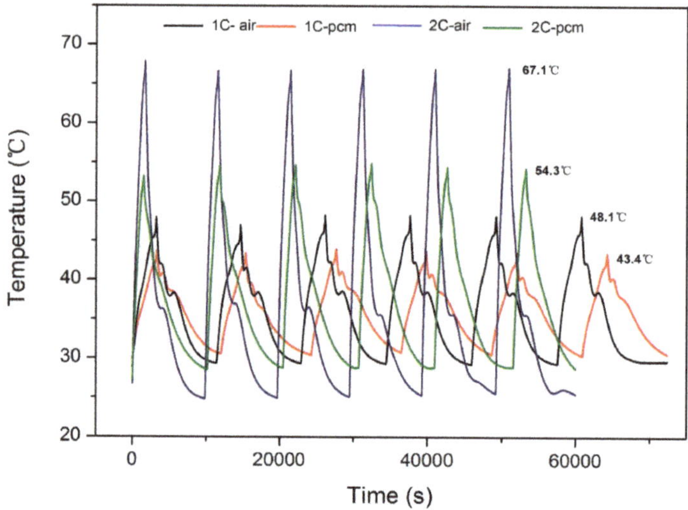

(**a**) The temperature curves of Sanyo LIB

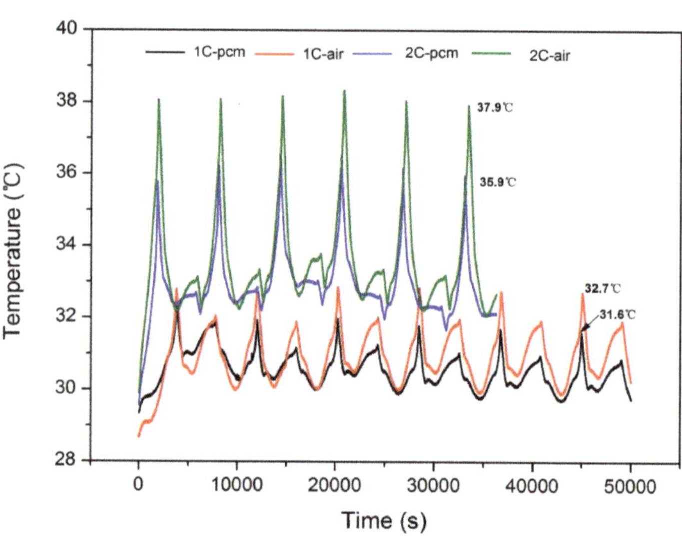

(**b**) The temperature curves of Sony LIB

Figure 7. The temperature curves of Sanyo and Sony LIB under 1 C, 2 C discharging.

4. Conclusions

PCM has been widely used in the thermal management of LIB due to itself advantages. In this paper, the changes of LIB temperature and capacity under different working conditions and discharge ratios are analyzed in order to further analyze the PCM influence of LIB thermal management and design an effective BTMS. The conclusions are as follows:

(1) The LIB temperature rises during the discharge process, and the PCM can effectively reduce the LIB temperature and keep its capacity stable. The average highest temperature of Sanyo LIB under the PCM cooling is about 54.4 °C in the 2 C discharge rate especially, and it is about 12.3 °C decrease compared with the natural convection cooling.

(2) The heat dissipation fins can reduce the LIB temperature, but the current effect is not very significant. Further optimization of the combination with heat dissipation fins and PCM is also an important direction in the thermal management of LIB.

(3) The LIB capacity changes are related to the temperature. In the discharge rate cycles, the LIB temperature of PCM cooling and PCM combined with heat dissipation fins decreased slightly, and the capacity decreases compared with the natural convection condition. The battery temperature and capacity increase slightly under extreme conditions.

(4) The PCM have a bigger temperature impact on Sanyo LIB. Sanyo LIB generates more heat due to its large capacity, and PCM absorbs more latent heat. The temperature reduction effect of PCM on high rate discharge LIB is more obvious.

In the future study, PCM combined with other cooling way efficiently and developing the PCM of higher latent heat, better heat conduction performance will be an important direction of battery BTMS.

Author Contributions: Conceptualization, M.C., S.Z., and J.W. (Jingwen Weng); Methodology, M.C., D.O. and J.W. (Jian Wang); Investigation, L.Z., S.Z., and G.W.; Writing-Review & Editing, M.C., S.Z., and X.W. All authors have read and agreed to the published version of the manuscript.

Funding: This research was funded by the National Key Research and Development Program of China (2018YFC0808600), the Open Project of State Key Laboratory of Fire Science (HZ2020-KF08), Natural Science Foundation of the Higher Education Institutions of Jiangsu Province (19KJB620003), and the Double Innovation Plan of Jiangsu province.

Conflicts of Interest: The authors declare no conflict of interest.

Nomenclature

Air	Air cooling
BTMS	Battery thermal management system
Excon	Extreme condition
LiFePO4	Lithium iron phosphate
LIB	Lithium-ion battery
PCM	Phase change material
PCM-fin	PCM with heat dissipation fins
SOC	State of charge

References

1. Li, X.; He, F.; Ma, L. Thermal management of cylindrical batteries investigated using wind tunnel testing and computational fluid dynamics simulation. *J. Power Sources* **2013**, *238*, 395–402. [CrossRef]
2. Weng, J.; Yang, X.; Ouyang, D.; Chen, M.; Zhang, G.; Wang, J. Comparative study on the transversal/lengthwise thermal failure propagation and heating position effect of lithium-ion batteries. *Appl. Energy* **2019**, *255*, 113761. [CrossRef]
3. Ouyang, D.; Weng, J.; Hu, J.; Liu, J.; Chen, M.; Huang, Q.; Wang, J. Effect of high temperature circumstance on lithium-ion battery and the application of phase change material. *J. Electrochem. Soc.* **2019**, *166*, A559–A567. [CrossRef]
4. Cicconi, P.; Landi, D.; Germani, M. Thermal analysis and simulation of Li-ion battery pack for a lightweight commercial EV. *Appl. Energy* **2017**, *192*, 159–177. [CrossRef]
5. Maleki, H.; Deng, G.; Anani, A.; Howard, J. Thermal stability studies of Li-ion cells and components. *J. Electrochem. Soc.* **1999**, *146*, 3224–3229. [CrossRef]

6. Qian, Z.; Li, Y.; Rao, Z. Thermal performance of lithium-ion battery thermal management system by using mini-channel cooling. *Energy Convers. Manag.* **2016**, *126*, 622–631. [CrossRef]
7. Duan, X.; Naterer, G.F. Heat transfer in phase change materials for thermal management of electric vehicle battery modules. *Int. J. Heat Mass Transf.* **2010**, *53*, 5176–5182. [CrossRef]
8. Zou, D.; Liu, X.; He, R.; Zhu, S.; Bao, J.; Guo, J.; Hu, Z.; Wang, B. Preparation of a novel composite phase change material (PCM) and its locally enhanced heat transfer for power battery module. *Energy Convers. Manag.* **2019**, *180*, 1196–1202. [CrossRef]
9. Chen, M.; Bai, F.; Song, W.; Lv, J.; Lin, S.; Feng, Z.; Li, Y.; Ding, Y. A multilayer electro-thermal model of pouch battery during normal discharge and internal short circuit process. *Appl. Therm. Eng.* **2017**, *120*, 506–516. [CrossRef]
10. Wang, Z.; Zhang, H.; Xia, X. Experimental investigation on the thermal behavior of cylindrical battery with composite paraffin and fin structure. *Int. J. Heat Mass Transf.* **2017**, *109*, 958–970. [CrossRef]
11. Esmaeili, J.; Jannesari, H. Developing heat source term including heat generation atrest condition for Lithium-ion battery pack by up scaling information from cell scale. *Energy Convers. Manag.* **2017**, *139*, 194–205. [CrossRef]
12. Pesaran, A.A. Battery thermal models for hybrid vehicle simulations. *J. Power Sources* **2002**, *110*, 377–382. [CrossRef]
13. Giuliano, R.M.; Prasad, A.K.; Advani, S.G. Experimental study of an air-cooled thermal management system for high capacity lithium titanate batteries. *J. Power Sources* **2012**, *216*, 345–352. [CrossRef]
14. Chen, D.; Jiang, J.; Kim, G.-H.; Yang, C.; Pesaran, A. Comparison of different cooling methods for lithium ion battery cells. *Appl. Therm. Eng.* **2016**, *94*, 846–854. [CrossRef]
15. Zhang, Y.; Yu, X.; Feng, Q.; Zhang, R. Thermal performance study of integrated cold plate with power module. *Appl. Therm. Eng.* **2009**, *29*, 3568–3573. [CrossRef]
16. Zhang, G.Q.; Zhang, H.Y. Progress in application of phase change materials in battery module thermal management system. *Mater. Rev.* **2006**, *20*, 9–12. (In Chinese)
17. Liang, Z.; Zhongke, S. Application progress of phase change energy storage technology in automobile energy saving. *J. Chem. Ind. Eng.* **2018**, *69*, 17–25.
18. Du, K.; Calautit, J.; Wang, Z.; Wu, Y.; Liu, H. A review of the applications of phase change materials in cooling, heating and power generation in different temperature ranges. *Appl. Energy* **2018**, *220*, 242–273. [CrossRef]
19. Rao, Z.; Wang, S. a review of power battery thermal energy management. *Renew Sustain. Energy Rev.* **2011**, *15*, 4554–4571. [CrossRef]
20. Kandasamy, R.; Wang, X.Q.; Mujumdar, A.S. Application of phase change materials in thermal management of electronics. *Appl. Therm. Eng.* **2007**, *27*, 2822–2832. [CrossRef]
21. Nayak, K.C.; Saha, S.K.; Srinivasan, K.; Dutta, P. A numerical model for heat sinks with phase change materials and thermal conductivity enhancers. *Int. J. Heat Mass Transfer* **2006**, *49*, 1833–1844. [CrossRef]
22. Duan, X.; Naterer, G.F. Thermal protection of a ground layer with phase change materials. *ASME J. Heat Transfer* **2010**, *132*, 011301. [CrossRef]
23. Al-Hallaj, S.; Selman, J.R. A novel thermal management system for EV batteries using phase change material (PCM). *J. Electrochem. Soc.* **2000**, *147*, 3231–3236. [CrossRef]
24. Al-Hallaj, S.; Selman, J.R. Novel Thermal Management of Battery Systems. U.S. Patent 6,468,689 B1, 22 October 2002.
25. Mills, A.; Al-Hallaj, S. Simulation of passive thermal management system for lithium-ion battery packs. *J. Power Sources* **2005**, *141*, 307–315. [CrossRef]
26. Javani, N.; Dincer, I.; Naterer, G.F. Numerical modeling of sub module heat transfers with phase change material for thermal management of electric vehicle battery packs. *J. Therm. Sc. Eng. Appl.* **2015**, *7*, 031005. [CrossRef]
27. Weng, J.; Yang, X.; Zhang, G.; Ouyang, D.; Chen, M.; Wang, J. Optimization of the detailed factors in a phase-change-material module for battery thermal management. *Int. J. Heat Mass Transf.* **2019**, *138*, 126–134. [CrossRef]
28. Javani, N.; Dincer, I.; Naterer, G.F.; Yilbas, B.S. Heat transfer and thermal management with PCMs in a Li-ion battery cell for electric vehicles. *Int. J. Heat Mass Transf.* **2014**, *72*, 690–703. [CrossRef]

29. Ling, Z.; Chen, J.; Fang, X. Experimental and numerical investigation of the application of phase change materials in a simulative power batteries thermal management system. *Appl. Energy* **2014**, *21*, 104–113. [CrossRef]
30. Wang, Z.; Zhang, Z.; Jia, L.; Yang, L. Paraffin and paraffin/aluminum foam composite phase change material heat storage experimental study based on thermal management of Li-ion battery. *Appl. Therm. Eng.* **2015**, *78*, 428–436. [CrossRef]
31. Li, W.Q.; Qu, Z.G.; He, Y.L.; Tao, Y.B. Experimental study of a passive thermal management system for high-powered lithium ion batteries using porous metal foam saturated with phase change materials. *J. Power Sources* **2014**, *255*, 9–15. [CrossRef]
32. Kizilel, R.; Sabbah, R.; Selman, J.R.; Al-Hallaj, S. an alternative cooling system to enhance the safety of Li-ion battery packs. *J. Power Sources* **2009**, *194*, 1105–1112. [CrossRef]
33. Zhao, J.; Lv, P.; Rao, Z. Experimental study on the thermal management performance of phase change material coupled with heat pipe for cylindrical power battery pack. *Exp. Therm. Fluid Sci.* **2017**, *82*, 182–188. [CrossRef]
34. Rao, Z.; Huo, Y.; Liu, X. Experimental investigation of battery thermal management system for electric vehicle based on paraffin/copper foam. *J. Energy Inst.* **2015**, *88*, 241–246. [CrossRef]
35. Alipanah, M.; Li, X. Numerical studies of lithium-ion battery thermal management systems using phase change materials and metal foams. *Int. J. Heat Mass Transf.* **2016**, *102*, 1159–1168. [CrossRef]
36. Hussain Tso, C.Y.; Chao, C.Y.H. Experimental investigation of a passive thermal management system for high-powered lithium ion batteries using nickel foam-paraffin composite. *Energy* **2016**, *115*, 209–218. [CrossRef]
37. Shang, S.; Jian-zu, Y.U.; Yong-qi, X.I.E.; Hong-xia, G.A.O.; ming, L. Temperature rise characteristics of phase change material/air cooling integrated thermal management system for lithium batteries. *J. Beijing Univ. Aeronaut. Astronaut.* **2017**, *43*, 1278–1286.
38. Azizi, S.M.Y.; Sadrameli, Y. Thermal management of a LiFePO4 battery pack at high temperature environment using a composite of phase change materials and aluminum wire mesh plates. *Energy Convers. Manag.* **2016**, *128*, 294–302. [CrossRef]
39. Sun, Z.; Fan, R.; Yan, F.; Zhou, T.; Zheng, N. Thermal management of the lithium-ion battery by the composite PCM-Fin structures. *Int. J. Heat Mass Transf.* **2019**, *145*, 118739. [CrossRef]
40. Zhang, Q.; White, R.E. Capacity fade analysis of a lithium ion cell. *J. Power Sources* **2008**, *179*, 793–798.

Article

Comparison of Heat Transfer Enhancement Techniques in Latent Heat Storage

William Delgado-Diaz, Anastasia Stamatiou *, Simon Maranda, Remo Waser and Jörg Worlitschek

Competence Center Thermal Energy Storage (CCTES), Lucerne University of Applied Sciences and Arts, 6048 Horw, Switzerland; williamorlando.delgadodiaz@hslu.ch (W.D.-D.); simon.maranda@hslu.ch (S.M.); remo.waser@hslu.ch (R.W.); joerg.worlitschek@hslu.ch (J.W.)

* Correspondence: anastasia.stamatiou@hslu.ch

Received: 22 June 2020; Accepted: 3 August 2020; Published: 10 August 2020

Abstract: Latent Heat Energy Storage (LHES) using Phase Change Materials (PCM) is considered a promising Thermal Energy Storage (TES) approach as it can allow for high levels of compactness, and execution of the charging and discharging processes at defined, constant temperature levels. These inherent characteristics make LHES particularly attractive for applications that profit from high energy density or precise temperature control. Many novel, promising heat exchanger designs and concepts have emerged as a way to circumvent heat transfer limitations of LHES. However, the extensive range of experimental conditions used to characterize these technologies in literature make it difficult to directly compare them as solutions for high thermal power applications. A methodology is presented that aims to enable the comparison of LHES designs with respect to their compactness and heat transfer performance even when largely disparate experimental data are available in literature. Thus, a pair of key performance indicators (KPI), Φ_{PCM} representing the compactness degree and NHTPC, the normalized heat transfer performance coefficient, are defined, which are minimally influenced by the utilized experimental conditions. The evaluation procedure is presented and applied on various LHES designs. The most promising designs are identified and discussed. The proposed evaluation method is expected to open new paths in the community of LHES research by allowing the leveled-ground contrast of technologies among different studies, and facilitating the evaluation and selection of the most suitable design for a specific application.

Keywords: heat transfer; high power; latent heat; energy storage; heat exchanger

1. Introduction

On the path to the integration of an ever-increasing share of variable renewable energy sources (VRES) into the current energy system, energy storage (ES) technologies play a fundamental role. Energy transformation and consumption globally account for more than 60% of the total green house gas emissions [1]. Additionally, in Switzerland and the European Union in general, over 50% of the total energy consumed is ultimately used as thermal energy for both industry and domestic applications [2,3]. Considering this, the development of thermal energy storage (TES) systems has become a priority for directly pure thermal applications and heat management systems, as well as combined electro-thermal storage initiatives, such as pumped thermal energy storage systems [4] and its potential for alternative use and flexibility for recovered waste heat from already existing sources at large scales [5].

Within the spectrum of TES technologies, Latent Heat Energy Storage (LHES) systems using Phase Change Materials (PCM) allow for thermal energy storage and release within narrow temperature differences with high energy density when compared to the sensible energy storage (SES) approach. These characteristics ultimately allow for the implementation of TES systems with a high degree of compactness. The heat transfer performance of LHES systems is however hindered by the time-dependent nature of its charging and discharging processes. During crystallization, the heat transfer and phase change processes are thus dominated by conduction under increasing resistance imposed by the moving liquid-solid front, making them a function of the state of charge of the unit [6]. Thus, the widespread application of LHES units on processes that require high heat transfer rate and quick response time relies on the many novel promising technological approaches that have emerged as a way to bypass LHES heat transfer limitations. These technologies include plain tube and finned tube bundle configurations [7–14], carbon composites and dispersions [15–20], metal foams [21–26], macroencapsulation techniques [27–31], and addition of conductive nanoparticles [32–34], among others.

The extensive range of experimental conditions (e.g., inlet temperature and mass flow rate of heat transfer fluid, phase change temperature, size of storage unit, etc.) used to characterize these technologies in literature make it difficult to directly compare and cross-correlate various performance features. This variability makes heat transfer performance and the degree of compactness especially hard to assess without leveled Key Performance Indicators (KPI). Some KPIs have been proposed to evaluate different aspects of an LHES unit. For instance, Energy density (ED) as proposed by Romani et al. [35] provides an indication of the amount of energy stored in relationship to the volume or mass of the unit. ED allows for valuable comparison of TES technologies in terms of overall capacity and required space and material resources, but most relevantly in the case of LHES, it equally considers sensible and latent contributions. By considering the sensible part of the energy stored in the PCM, the result is dependent on the operation temperature levels on the unit and not only the materials and amounts. Alternatively, the energy efficiency ratio described by Wang et al. [36] concerns the ratio of energy required to pump the HTF through the LHES unit to the stored energy. In addition, in a similar approach, Li et al. [37] also proposed the performance analysis of a wide range of LHES operational and material parameters, as both energetic and exergetic efficiencies. Even though the previously described KPIs provide very valuable information about an LHES unit, they address particular aspects and consider mostly capacity and efficiency perspectives, but provide no indication on the rates of heat transfer and the required material to achieve it.

Directly addressing the thermal response, Gasia et al. [38] proposed various KPI for both short and long term scales: Average power, 5 min-peak-power, 5 min peak power-energy ratio (based on 5-min-peak-power over total capacity of the unit), and finally the discharge time. The set of KPIs was intended for the evaluation of four LHES units of very similar scales, operating at uniform conditions and equal PCM. Although useful while the experimental conditions and the geometry remain similar enough, they remain intrinsically connected to the current operation conditions and scale. This leads to non-representative results, especially when comparing across different studies and applications. Similarly, Guo et al. [39] consider the specific charging rate ($\gamma[1/\mathrm{h}]$) and specific energy loss rate. The specific charging rate directly addresses the heat transfer performance of the unit, but it is ultimately an average power to capacity ratio, without any normalization with respect to the driving forces.

Similar analyses include, for instance, the use of the average temperature effectiveness (ε_{avg}) by Nomura et al. [40] and Krimmel et al. [41] to represent the efficiency of the heat exchange. Additionally, Nomura et al. [40] present a NHTPC (h_v), directly addressing the heat transfer performance of the units. It considers the average heat transfer rate divided by the average temperature difference (which ultimately can be interpreted as the enthalpy flow of the HTF), with respect to the volume of PCM only in the unit. Although it directly addresses the heat transfer capability of a unit, it still remains dependent on the HTF conditions, and thus varies with different mass flow and temperature difference.

The Effectiveness-NTU Method (ε-NTU) allows the calculation of the heat transfer rate and temperature profiles in a heat exchanger using the enthalpy flows, and defining a heat exchanger effectiveness based on the actual heat transfer rate over the maximum achievable by the system [42]. It has been previously used to analyze LHES systems as performed by Tay et al. [43,44] as design and sizing tool on specific designs.

The methodology presented in this paper is inspired on the previous (ε-NTU) analyses and the view of LHES units as a heat exchanger core acting as boiler or condenser. It focuses on two KPIs, which represent the normalized heat transfer performance and degree of compactness of the LHES design. This new approach allows the comparison of LHES systems reported in literature in terms of their heat transfer capabilities and compactness regardless of their geometry, scale, and operation conditions. The developed KPIs are applied to several technologies reported in literature and the results of this comparison are presented and discussed. Based on the authors' knowledge, this is the first time such an extensive quantitative comparison across different LHES technologies with a focus on high power applications has been performed.

2. Methodology

Definition of Proposed KPIs

The main focus of this study is to allow the simplified and quick evaluation of the heat transfer capabilities in a LHES unit regardless of scale and operating conditions. Achieving this goal requires the usage of the most readily available information able to represent a highly transient process through averaged properties. The methodology proposed in this study uses this information adapted around the (ε-NTU) method. Two KPI are proposed representing both the heat transfer performance of a LHES unit as well as the degree of compactness and an indication of the energy density attainable by the system.

Regarding the degree of compactness, the volume fraction of the major contributor to the storage capacity of the LHES unit, PCM to total volume of the unit, (Φ_{PCM}) is suggested as an indicator of the energy density attainable by the system. Φ_{PCM} provides an indication of the compactness degree attainable, but also information on the required trade-off of PCM storage volume for heat exchanger material to achieve certain heat transfer performance, and it is the ratio of PCM volume in the storage (V_{PCM}) to the total outer volume of the unit (V_{TOT}). See Equation (1)

$$\Phi_{PCM} = \frac{V_{PCM}}{V_{TOT}} \tag{1}$$

The total volume (V_{TOT}) was calculated considering the geometry of the outermost layer of the unit while excluding the additional volume used for insulation. The container wall thickness as well as additional volume dedicated to manifolds (flow development) and the like were all taken into account.

From a heat exchanger perspective, an LHES unit can be regarded as a heat exchanger, the performance of which is defined by a static heat sink or source, or an analog case of a heat exchanger operating as a boiler or condenser. The average NTU (NTU_{avg}) represents the added effects of the heat exchanger tubes and growing layer of solid through the discharge process. It is defined as the ratio of the heat transfer rate capacity of the heat exchanger (product of the overall heat transfer coefficient (U) and the heat exchange surface (A)), and the heat capacity rate of the HTF ($\dot{m}_{HTF} \cdot c_{p,HTF}$) [45].

Additionally, the average NTU can be easily estimated as it is directly related to the average effectiveness (ε_{avg}) of the heat exchange, and under the assumptions of a phase change, similarly to boiler/condenser operation, the heat exchanger effectiveness relations can be simplified [45] as described in Equation (2):

$$NTU_{avg} = \frac{U \cdot A}{\dot{m}_{HTF} \cdot c_{p,HTF}} = -ln(1 - \varepsilon_{avg}) \tag{2}$$

This equation can be rearranged and divided by the total volume V_{TOT} to calculate the normalized heat transfer performance coefficient (NHTPC) as shown below in Equation (3):

$$\mathbf{NHTPC} = \frac{\mathbf{U \cdot A}}{\mathbf{V}_{TOT}} = \frac{-ln(1-\varepsilon_{avg})\cdot \dot{m}_{HTF}\cdot c_{p,HTF}}{V_{TOT}} \tag{3}$$

Thus, the proposed KPI on the heat transfer side, NHTPC, can be seen as the product of the overall heat transfer coefficient (U) and the heat exchange surface (A) normalized by the total volume of the LHES unit (V_{TOT}). Considering the transient nature of the solidification process due to the increasing resistance and changing surface, it is useful to express an average U·A for the whole process. Dividing this product by the total volume of the LHES unit (V_{TOT}) excluding insulation enables comparison of the heat transfer behavior regardless of the final dimensions, operation conditions, and overall scale.

Where ε_{avg} represents the average heat exchanger effectiveness during discharge, $\dot{m}_{HTF}$ and $c_{p,HTF}$, the mass flow rate and specific heat capacity of the HTF, respectively, and finally V_{TOT} the total outer volume of the container without considering any insulation.

In this case, the effectiveness of the heat transfer (ε_{avg}) is defined by the relation of the actual average (over the discharge time) temperature difference between inlet ($\overline{T_{HTF,In}}$) and outlet ($\overline{T_{HTF,Out}}$), and the theoretical maximum temperature difference achievable, with respect to the phase change temperature (T_{PC}) as shown in Equation (4) [43,45].

The T_{PC} values were reported by the individual studies, and are usually obtained through differential scanning calorimetry (DSC) measurements:

$$\varepsilon_{avg} = \frac{\overline{T_{HTF,In}} - \overline{T_{HTF,Out}}}{\overline{T_{HTF,In}} - T_{PC}} \tag{4}$$

This relation allows access to the average product "U·A", or the average heat rate capacity of the heat exchange geometry in the core of the unit. This quantity can be considered independent of the operation conditions, but remains an intrinsic characteristic of the heat exchanger geometry, design, and material combination (PCM and heat exchanger materials).

During the solidification process, the conductive resistance between the HTF and the solidifying (liquid) PCM increases as solid PCM builds up around the HEX structure surface. This also generates a changing solid–liquid PCM heat transfer surface throughout the process. With this in mind, it can be especially handy to consider the product U·A averaged through time, since both the heat transfer surface, and heat transfer coefficient, vary throughout the discharge process with the state of charge of the unit.

Even though the main required information pertaining to the geometry, materials, and amounts are uniformly available, how and which experimental results are readily displayed in literature remains very dependent on the authors and the focus of the studies. Some additional considerations to the definition of the pair of the previously discussed KPIs are suggested for an even representation with the proposed KPIs:

The average HTF outlet temperature ($\overline{T_{HTF,Out}}$) was preferably estimated by fitting a polynomial function to the reported data and calculating a mean function value between the beginning of the discharge process up to an arbitrary point. For practical purposes, and considering that once a high degree of solidification is attained the power sharply decreases, a 90% solidification or melting is considered as a standard for a completed process and thus is defined as discharge time (t_{Disch}).

For the few cases in which outlet temperature or power profiles were not provided, $\overline{T_{HTF,Out}}$ can be approximated as shown in Equation (5) derived from the simplified steady-flow thermal energy equation [45]:

$$\overline{T_{HTF,Out}} = \overline{T_{HTF,In}} + \frac{\dot{Q}_{avg}}{\dot{m}_{HTF}\cdot c_{p,HTF}} \tag{5}$$

Knowing t_{Disch} conveniently allows for indirectly representing the average output power-to-capacity ratio as it represents the inverse of the time required to achieve a certain amount of PCM solidification, and thus the average discharge power ($\dot{Q}_{avg}$) can be approximated using Equation (6) [46] and the energy balance based on material properties and temperature levels:

$$\frac{\dot{Q}_{avg}}{E_{st,90}} = \frac{1}{t_{Disch.}} \tag{6}$$

The energy associated with the defined standard degree of solidification ($E_{st,90}$) is estimated assuming the complete contribution of the sensible heat from container ($m_{Cont} \cdot c_{p,Cont}$) and heat exchanger materials ($m_{HEX} \cdot c_{p,HEX}$) and PCM ($m_{PCM} \cdot c_{p,PCM}$) from the initial temperature of the unit T_{init} up to the phase change temperature of the PCM T_{PC} in addition, to 90% of the latent contribution from the PCM, as shown in Equation (7):

$$E_{st,90} = (m_{HEX} \cdot c_{p,HEX} + m_{Cont} \cdot c_{p,Cont} + m_{PCM} \cdot c_{p,PCM}) \cdot (T_{init} - T_{PC}) + (90\% \cdot m_{PCM} \cdot \Delta h_{PC}) \tag{7}$$

Using these considerations and assumptions allows for the calculation of the proposed NHTPC with minimal representative information. The results of the preliminary analysis are shown and discussed subdivided in similarity classes, with a specific focus on the heat transfer performance and the potential of the different approaches for applications that require high power LHES systems.

In summary, Φ_{PCM} represents the ratio of main energy storage material to the total volume of the unit, excluding insulation. It provides a general idea on the compactness degree of the system and energy density potential by remaining independent of the temperature levels. Additionally, it can be regarded as a representation of the required HEX material to attain a certain heat transfer performance. It only requires the overall dimensions of the unit and the total amount of PCM inside for its calculation.

NHTPC represents the average heat transfer performance of an LHES unit regardless of the operation conditions (HTF flowrate and temperature levels) used by the authors of the different studies but remaining an intrinsic characteristics of the heat exchange structure geometry, and material combinations (PCM, HEX, container, etc). The calculation requires, in principle, information on both inlet and outlet temperatures on the unit and material properties of PCM and the different components (HEX and container). Temperature profiles are preferably used to estimate directly the average temperatures by using fitting techniques, mean function values, and the previously defined $E_{st,90}$ from the energy balance. Alternatively, if this information is not presented directly, average discharge power or the discharge time can be used to compute close approximations, as shown in Equations (5)–(7).

3. Results and Discussion

Only the studies that provided sufficient information to perform the calculations with minimal assumptions are shown and discussed in this study. The analysis of the different cases is presented subdivided in four subclasses, namely, finned tube bundle heat exchanger structures, composites of different natures as Thermal Conductivity Enhancers (TCE), macro encapsulation based systems, and experiments using automotive heat exchanger structures and capillary tube bundles.

3.1. Robustness Testing

In order to corroborate the relative independence of the proposed KPIs with regard to the operation conditions and scale of the LHES unit under scrutiny, a sensitivity analysis incorporating results from a sample of studies in which different inlet temperatures ($T_{HTF,In}$) and HTF mass flow rate ($\dot{m}_{HTF}$) were used.

The influence of the inlet temperature on the final NHTPC was examined using data from Waser et al. [7] and considering three $T_{HTF,In}$ levels between 15 °C and 40 °C in both a tube bundle

(Unit 1) with 0.02 m^3 (20 kg of PCM, $CH_3COONa \cdot 3H_2O$), and the equivalent finned tube bundle (Unit 2) under a constant 360 kg/h of flowing water as HTF. The mass flow rate effect was considered using data from two different sources and consistent temperature levels. On a larger scale, the data gathered by Zauner et al. [47] for a storage of 0.4 m^3 total outer volume (170 kg of PCM, HDPE) operating with thermal oil (Marlotherm SH) as HTF between 2088 kg/h up to 6984 kg/h (Unit 3) was used. On the smaller side, the unit shown by Amagour et al. [13] is considered (Unit 4) with a total outer volume of 0.009 m^3 (2.3 kg of PCM, organic blend) with water as HTF and flow rates ranging between 24 kg/h and 62 kg/h.

Figure 1 shows the obtained results for the analysis of the ability of the KPIs to describe system performance independently of experimental conditions.

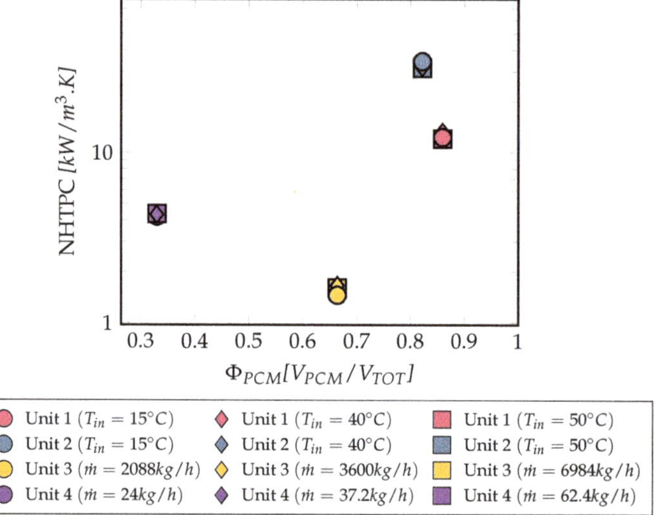

Figure 1. Sensitivity analysis. Calculated NHTPC vs. Φ_{PCM} at different temperature levels and HTF mass flow rates in logarithmic scale.

The tight spread of the final results for all four cases corroborates the relative independence of the proposed KPI to the operation conditions. Even though small variations were found for the different experiments, an average coefficient of variation (defined as the standard deviation over the average value) of less than 6% was calculated for every case, which is considered satisfactory for the purpose of this study.

3.2. Finned Tube Bundles

Finned Tube Bundle (FTB) geometries are the most widely studied Heat Exchanger (HEX) configuration for LHES applications. They consist of fixed-fin geometries, of different topologies and configurations, generally protruding from the tubes used to carry the HTF, into the PCM mass. Highly conductive metals such as copper and aluminum are the most commonly used. The disposition of the fins, packing fraction, thickness, and amount determine the performance of a finned tube bundle.

The selection of analyzed studies which addressed the addition of fins to a tube bundle (TB) heat exchanger geometry to enhance the heat transfer rates of a LHES unit is summarized in Table 1. It includes reference numbers, notation, some descriptive information, and a representative picture or scheme of the discussed units. All additional data required for the calculation of NHTPC are presented in Table A1. The outer volumes of the units V_{TOT}, including the container tank surrounding the TB or FTB (and additional manifolds if required) without considering any insulation, were for the most part explicitly reported along with container material and properties. In the cases were information was missing in this regard, it could be approximated from additional reported geometrical parameters and conservative assumptions. Figure 2 presents the estimated performance indicators. Experiments performed in the framework of the same study with different geometries are indicated with the same letter but different numbering.

Table 1. References pertaining to tube bundles (TB) and finned tube bundles (FTB).

Type	T_{PC} [°C]	V_{TOT} [m^3]	$\dot{m}_{HTF}$ [kg h^{-1}]	$c_{p,HTF}$ [kJ/(kg · K)]	Geometry	Ref.
TB (A.1)	58	2.0×10^{-2}	3.6×10^{2}	4.18	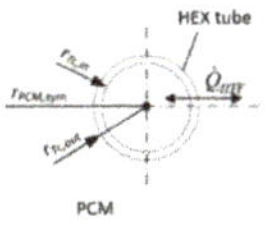	[7]
FTB (A.2)	58	2.0×10^{-2}	3.6×10^{2}	4.18	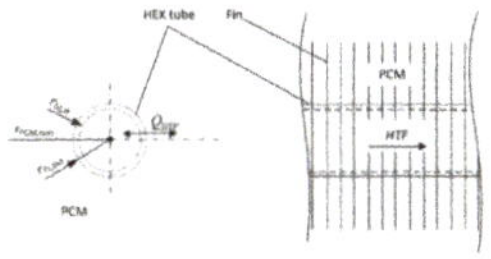	[7]
TB (B.1)	35	9.3×10^{-4}	4.0×10^{2}	4.18	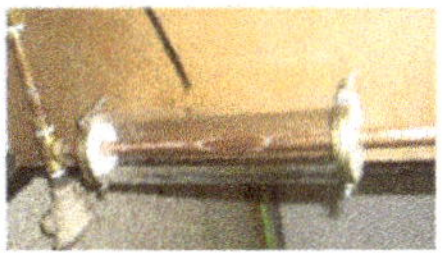	[48]
FTB (B.2)	35	9.3×10^{-4}	4.0×10^{2}	4.18	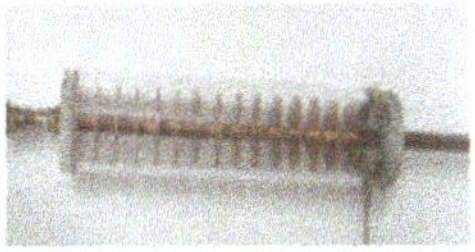	[48]
Triplex FTB(C)	82	1.6×10^{-2}	2.4×10^{2}	4.18	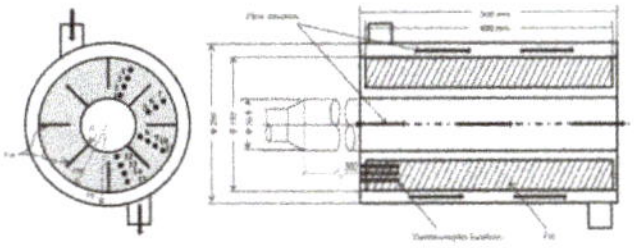	[9]
FTB (D.1)	94	3.4×10^{-3}	2.2×10^{3}	4.06	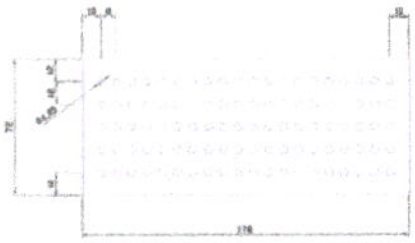	[14]
FTB (E)	305	3.5×10^{-1}	8.1×10^{3}	2.30		[10]
FTB (F)	142	8.7×10^{-2}	2.4×10^{2}	2.49	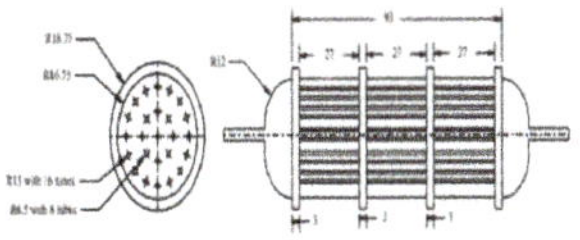	[8]

Table 1. *Cont.*

Type	T_{PC}[°C]	V_{TOT} [m^3]	$\dot{m}_{HTF}$ [kg h^{-1}]	$c_{p,HTF}$ [kJ/(kg·K)]	Geometry	Ref.
FTB (G)	125	3.9×10^{-1}	1.5×10^{3}	2.03	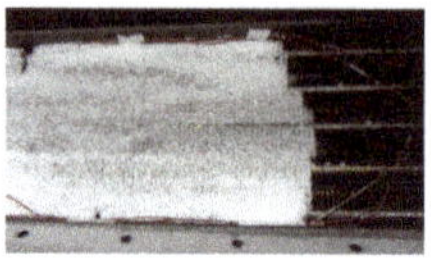	[47]
FTB (H)	53	8.7×10^{-3}	2.4×10^{1}	4.18		[13]
FTB (I)	42	6.1×10^{-2}	1.8×10^{2}	4.18	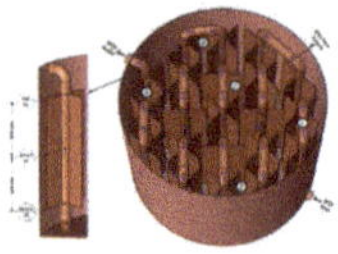	[49]
FTB (J)	169	6.5×10^{-3}	3.5×10^{2}	3.10	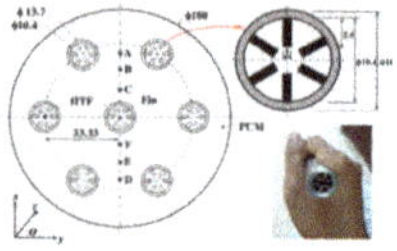	[12]

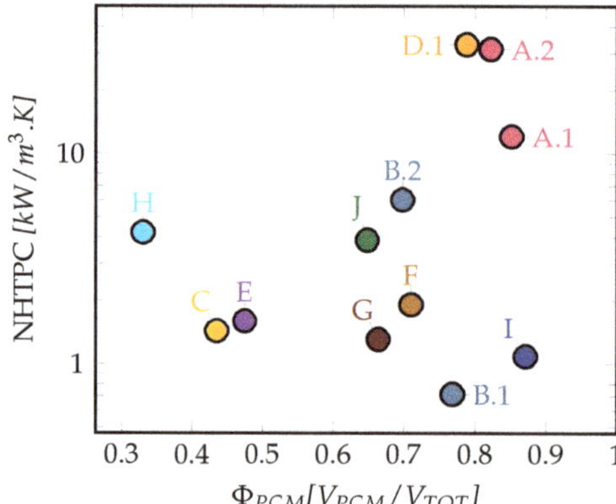

Figure 2. Calculated NHTPC vs. Φ_{PCM} for the references concerning tube bundles (TB) and finned tube bundles (FTB) in LHES unit configurations in logarithmic scale.

The most commonly found arrangements pertain fixed fins embedded in the PCM bulk, oriented perpendicularly (A.2, B.1, D, G, H) or longitudinally (C, E, F, I, J) to the HTF flow inside the tubes. Some interesting alternative variations see longitudinal fins used such as in triplex heat exchanger configurations by Al-Abidi et al. [9] (C), complex geometries to fit specific applications by Laing et al. [10] (E) or even fixed on the side of the flowing HTF, as presented by Raul et al. [12] (J). Regarding performance, the study carried out by Waser et al. [7] provided data for a finned copper tube bundle (A.2) and additional in-house data from the same study was used for the analog plain tube bundle (A.1). The addition of aluminum fins traded an additional 3% of PCM volume fraction for a significant discharge time reduction of around 50% and a total threefold improvement in the calculated NHTPC. However, it is necessary to highlight the performance of the tube bundle structure (A.1) as it exceeds most of the FTB designs. Taking the unit used by Khan et al. [49] (I) as comparison point, the sheer difference in heat transfer surface could explain this behavior. Even though A.1 presents no fins, the heat transfer surface to volume ratio of unit A.1 is over two times higher (63 m^2/m^3) than that of unit I (27 m^2/m^3).

In a similar comparison approach, Medrano et al. [48] (B.2) achieved a sevenfold increase (when compared to B.1) in heat transfer performance by trading an approximate 7% of PCM volume to accommodate circular radial fins.

The FTB unit presented by Shon et al. [14] (D.1) attained the higher performance in this case with an additional 4% of the total volume dedicated to the analog copper finned tubes than the FTB unit investigated by Waser et al. [7]. The heat exchange structure with a higher packing fraction could explain the differences in performance.

In the case of Amagour et al. [13] (H), it is interesting to note that around 65% of the total LHES unit volume was dedicated to heat exchange elements as well as additional empty space above the heat exchanger. It directly affects the overall energy density of the unit with around 29 kWh/m^3, but achieves a thermal response around the average for the category. This is evidence of the intricacies of the container, heat exchanger design, and chosen materials. A high share of heat exchanger material, or consequently a low PCM fraction, does not necessarily translate into improved heat transfer behavior. A similar situation was found on some other cases leading to a low Φ_{PCM} was caused by a large volume occupied by the HTF, and additional head space in the units which could potentially be optimized.

3.3. Automotive Heat Exchangers and Polymer Based Capillary Scale Tube Bundles

This entire subsection is dedicated to particularly interesting systems as they are in principle extreme variations of finned and simple tube bundles. On one hand, Automotive HEX (AHEX), conceived from the idea of finned tubes for gas-to-liquid heat exchange, are mass produced on a wide variety of configurations usually completely made out of braced aluminum and seek to maximize the heat transfer surface. On the other hand, without additional material, very large heat transfer surfaces can be achieved by driving the diameter of a tube bundle to the capillary scale (below 5 mm). Additionally, the consequent thin walls produced at this scale minimize the influence of the material conductivity, opening the door to the utilization of polymers for a wide range of applications when temperature levels allow. These systems will be finally referred to as capillary tube bundles (CTB). Table 2 gathers the notation and representative schemes of the studies treated in this section. All additional data required for the calculation of NHTPC are presented in Table A1. The outer volumes of the units V_{TOT} , including the container tank surrounding the HEX without considering any insulation, were for the most part explicitly reported in the different studies along with container material and properties. In the cases where information was missing in this regard, it could be approximated from additional reported geometrical parameters and conservative assumptions.

The summarized KPIs are portrayed in Figure 3. Experiments performed in the framework of the same study with different geometries are indicated with the same letter but different numbering.

Medrano et al. [48] experimented with an AHEX unit embedded in PCM (B.4), and compared it to other common LHES approaches and it achieved an impressive NHTPC, several times higher than the next best performance within the same study, the graphite matrix enhanced tube-and-shell (B.3) unit previously discussed. This effect might be explained mainly by the sheer mass of heat exchanger material in the unit rather than the heat exchanger design as it presents a relatively low performance in both heat transfer and compactness when compared to other AHEX-based units.

Table 2. References pertaining to AHEX and CTB based units.

Type	T_{PC}[°C]	V_{TOT} [m^3]	$\dot{m}_{HTF}$ [$kg\,h^{-1}$]	$c_{p,HTF}$ [$kJ/(kg \cdot K)$]	Geometry	Ref.
AHEX (B.4)	35	4.4×10^{-3}	4.00×10^{2}	4.18	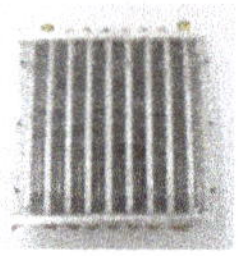	[48]
AHEX (S)	4	4.1×10^{-3}	5.34×10^{2}	1.01	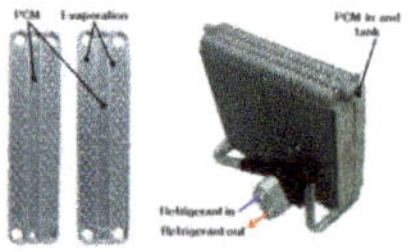	[50]
AHEX(T)	69.3	3.3×10^{-3}	7.13×10^{2}	3.95		[51]
AHEX(D.2)	93.8	1.9×10^{-3}	2.21×10^{3}	4.06		[14]
CTB (U)	30	1.0	3.60×10^{3}	4.18	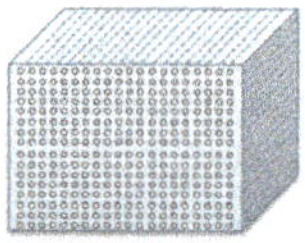	[52]
CTB (V)	29	8.5×10^{-1}	1.38×10^{3}	4.18		[53]
CTB (W)	28.5	1.5	3.15×10^{3}	4.18		[54]

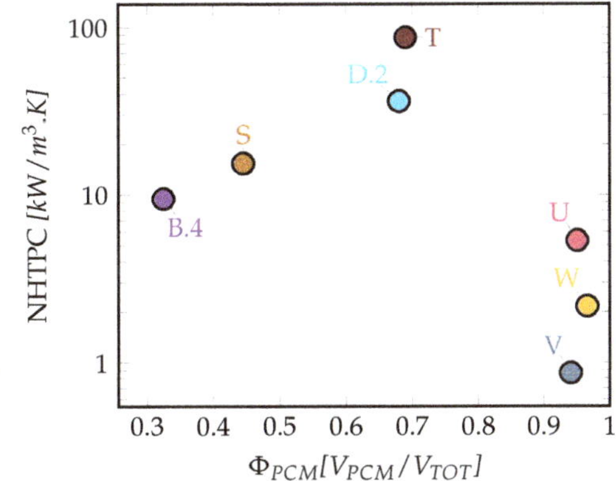

Figure 3. Calculated NHTPC vs. Φ_{PCM} for the references concerning AHEX and CTB in LHES unit configurations in logarithmic scale.

A similar situation was discussed in the study performed by Shon et al. [14] their initial experiments were performed using a stock automotive HEX immersed in PCM (D.2). Based on their results, an FTB design with a higher capacity was produced and is discussed in Section 3.2 (D.1). Interestingly, their goal was achieved as their FTB custom design (33 [kW/m^3·K]) effectively matched the thermal performance of the AHEX-based unit (36 [kW/m^3·K]). Their design variations, however, required an additional sacrifice of 10% PCM volume fraction. A second iteration of an analog system for a diesel based system, explored by Park et al. [51] (T), produced a custom heat exchanger with an analog flat pipe and fin configuration, and produced the highest NHTPC reported, reaching around 85 [kW/m^3·K] with a similar PCM volume fraction as its predecessor. The LHES unit proposed by Lee et al. [50] (S) constitutes a particular example of how LHES capacity and power are tailored for specific applications. The high performance parameter calculated and the significantly low PCM volume fraction were adjusted to the envisioned application requirements. The unit was manufactured to produce high cooling power for very short periods of time while the car is idle on a red light, as the LHES unit is part of an automotive HVAC system, and is thus adapted to be used with two HTF systems, coolant loops and pure air convection.

These are good examples of the potential of AHEX as highly optimized systems readily available at industrial manufacturing scales in a broad variety of configurations that, although envisioned for a different application, could widen the areas of implementation of an already existing product.

When looking at polymer CTBs, their average performance does not deviate much from the results seen for FTB. They are polymer based and large heat exchange surfaces are achievable while requiring very low volume within the unit, in the order of 90% PCM fraction or more. Helm et al. produced a prototype [53] (V) and posterior improvement [54] (W) as part of a solar heating and cooling system. The heat exchange structure based on polypropylene capillary tubes in the order of 4.3 mm outer diameter was used in both cases and not only its performance, but also its durability, were put to the test in system level experiments and cycling tests.

Similarly, Hejcik et al. [52] (U) presents a special case, studying the use of polypropylene hollow fibers, capillary tubes in the order of 0.8 mm outer diameter as tube bundle arrangements embedded in PCM within a modeling study. The reported performance was calculated based on the simulations carried out by the authors and the PCM volume fraction was calculated based on the model domain used which included only a PCM mass and the hollow fiber bundle. The potential of hollow fiber bundles and, in general, polymer capillary scale systems becomes discernible when contemplating that thermal performances comparable to FTB configurations are attainable using basic geometries that occupy a minimal share of the volume.

Even though the heat transfer performance seems adequate from a KPI perspective, it is necessary to clarify that although a quick discharge can be achieved with CTB based systems, they require special attention in their design as achieving a stable temperature output window requires low mass flow rates, long tubes, or systems in series. The added frictional effects of the HTF flow at low inner diameter conditions must be considered during the optimization to avoid excessive pumping power requirements and affecting the effectiveness of the system.

3.4. Composites

The use of highly conductive materials to increase the performance of PCM as Thermal Conductivity Enhancements (TCE) has been widely studied. The general goal of this kind of TCE is to allow the creation of a conductive network through the PCM mass and enhance overall conduction in both melting and solidification processes. Table 3 shows the notation and representative figures of the studies considered in this subsection. All additional data required for the calculation of NHTPC are presented in Table A1. The outer volumes of the units V_{TOT}, including the container tank surrounding the HEX (and additional manifolds if required) without considering any insulation, were for the most part explicitly reported in the different studies along with container material and properties. In the cases where information was missing in this regard, it could be approximated from additional reported geometrical parameters and conservative assumptions.

Table 3. References pertaining to carbon based techniques and metal foams as TCE

Type	T_{PC}[°C]	V_{TOT} [m^3]	$\dot{m}_{HTF}$ [kg h^{-1}]	$c_{p,HTF}$ [kJ/(kg · K)]	Geometry	Ref.
Graphite matrix (B.3)	35	9.3×10^{-4}	400	4.18		[48]
TB (K.1); Carbon brushes (K.2)	49	2.0×10^{-2}	32.6	4.18		[55]
Carbon cloth (L)	49	2.0×10^{-2}	32.6	4.18		[19]
Graphite matrix (M)	53	1.6×10^{-2}	30	4.18		[16]
TB (N.1); 95% FP (N.2); 77% FP (N.3)	58	6.2×10^{-4}	8.14	1.01		[22]

Figure 4 summarizes the performance results of carbon and metal based TCE in different configurations, such as carbon dispersions and composites of various forms, and metal foams of different amounts of pores per inch. Each subcategory will be further discussed separately.

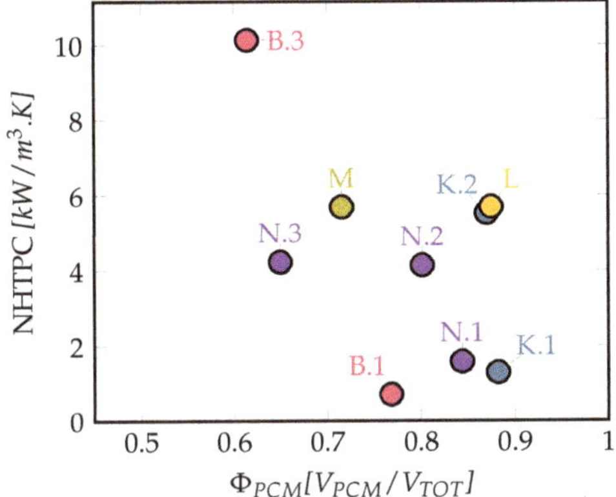

Figure 4. Calculated NHTPC vs. Φ_{PCM} for the references concerning carbon based structures and metal foams as TCE in LHES unit configurations.

3.4.1. Metal Foam Based Composites

Across the metal foams (MF) considered as TCE options for LHES systems, the most widely studied are aluminum [21,22], copper [23–25], and nickel [26] foams.

For instance, Atal et al. [22] considered aluminum foams of different porosities (N.2, N.3) on a shell and tube arrangement and achieved reductions in discharge time of up to 63% and a proportional increase in its heat transfer performance when compared to the case with only PCM (N.1). However, only a marginal increase in performance is observed with decreasing metal foam porosity (FP) (and consequently PCM volume fraction) as shown in Figure 4. An ultimate difference of around 15% additional PCM volume fraction is sacrificed to accommodate the lower porosity foam but little to no effect is shown in terms of heat transfer enhancement. Additionally, the energy density of the system is heavily affected, decreasing from 112 kWh/m^3 in the case of pure PCM (N.1), to 108 kWh/m^3 for the 95% porosity foam (N.2) and finally 94 kWh/m^3 in the case of the 77% porosity foam (N.3). It becomes clear that the energy density trade-off when accomodating the higher porosity foam (N.2) is worth it in terms of performance, but, once a sufficient highly conductive network is created, there is no substantial enhancement in increasing the amount of metal in the unit.

Lazzarin et al. [21] also studied the effect of aluminum foams with slightly different porosities and number of pores per inch (PPI) achieving in the best case a reduction of around 90% on the solidification time. In a similar way, Mancin et al. [23] used copper foams of increasing PPI and attained a reduction in charging time of around 27%. Similarly, Xiao et al. [24] used copper and nickel foams of different amounts of PPI to improve PCM conductivity. From the measurements performed by the authors, a great improvement is evident for all cases when compared to the original 0.305 W/m·K. For instance, the copper foams embedded in the PCM produced conductivities of 5 W/m·K and 16 W/m·K for 97% and 88.9% porosity samples, respectively.

The latter examples, although worth mentioning, are not shown in this study since their experimental setup was not conceived as LHES units working with a heat transfer fluid and thus it was not possible to accurately calculate the proposed KPIs without inaccurate assumptions. Similarly, a major share of the considered references pertaining the use of metallic foams concerned effective thermal conductivity measurements, and experimental setups focused towards electronic heat management strategies. In order to compare in terms of the NHTPC methodology, either more experimental work or validated models that place these materials in a LHES unit configuration are necessary to further study and evaluate their potential.

3.4.2. Carbon-Based Composites

Among the considered references, several of them contained some form of carbon based TCE, in the form of expanded graphite composites and dispersions [15,16,20,48], carbon fiber (CF) [17], carbon foam [18], and even carbon fiber cloth (CC) [19].

Within the selected units, Medrano et al. [48] proposed the highest performance of the carbon based enhancements by placing and expanded graphite and PCM composite in a double pipe heat exchanger configuration (B.3) and compared it to a direct analog unit containing only PCM (B.1) discussed in Section 3.2. The addition of the graphite matrix required the trade of 15% in PCM volume fraction but achieved a seventeen fold increase in terms of NHTPC.

Fukai et al. [19,55] used carbon fiber brushes [55] (K.2) and carbon fiber cloth [19] (L) threaded around a copper tube bundle structure (K.1). Interestingly, both enhancements, under the same experimental conditions, reduced the discharge time by up to 45% by trading a minor share of the volume to accommodate the brushes [55] (K.2) and carbon cloth [19] (L) in both cases. Wu et al. [16] (M) produced a similar performance using shape stabilized 75:25 PCM and expanded graphite composite with a copper tube bundle configuration as heat transfer elements but required an additional 17% PCM volume fraction to achieve it.

3.5. *Macro-Encapsulation Solutions*

The considered references pertaining macro encapsulation techniques include both high and low temperature applications. The considered examples are shown in Table 4. All additional data required for the calculation of NHTPC are presented in Table A1. The outer volumes of the units V_{TOT}, including the container tank surrounding the bed of capsules (and additional manifolds if required) without considering any insulation, were for the most part explicitly reported in the different studies along with container material and properties. In the cases where information was missing in this regard, it could be approximated from additional reported geometrical parameters and conservative assumptions. The results are available in Figure 5.

Ma et al. [27] (O) and Wickramaratne et al. [28] (P) both presented LHES units using stainless steel encapsulation methods for high temperature applications on a range of temperatures around 550 °C. Even though the systems are relatively similar, the difference in performance could be explained in part by the fact that Ma et al. [27] (O) uses Al:Si alloy as PCM and Wickramaratne et al. [28] (P) proposed a eutectic salt mixture. This means that, besides having a much larger PCM phase change enthalpy (432 kJ/kg), in the first case, the most common drawback associated with LHES is minimized by the high PCM thermal conductivity in both phases. This ultimately translates into higher average power, and it is taken into account in the NHTPC, as the average outlet HTF temperature is much closer to the phase change temperature.

Table 4. References pertaining to macro encapsulation techniques

Type	T_{PC}[°C]	V_{TOT} [m^3]	$\dot{m}_{HTF}$ [kg h^{-1}]	$c_{p,HTF}$ [kJ/(kg · K)]	Geometry	Ref.
ME (O)	577	4.2×10^{-3}	19.8	1.10		[27]
ME (P)	515	1.3×10^{-2}	70.5	1.07		[28]
ME (Q)	60	4.7×10^{-2}	120	4.18		[30]
Pouch (R.1), Sphere (R.2)	58	2.7×10^{-1}	900	4.18		[31]

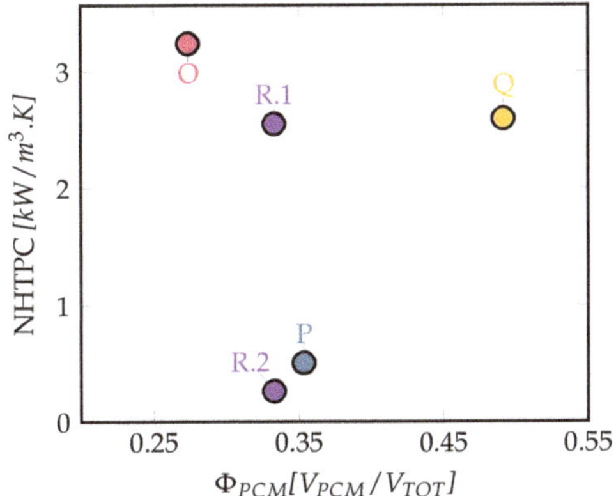

Figure 5. Calculated NHTPC vs. Φ_{PCM} for the references concerning macro encapsulation techniques in LHES unit configurations.

Park et al. [31] presented flexible (Polyethylene, Nylon, Aluminum, and PET) laminated pouches (R.1) as the encapsulation method and compared them to 3D CFD modeled spherical equivalents (R.2). The flexible pouches decreased the discharge time of the system by 62% when compared to the simulated spherical containers, and is reflected on a fivefold increase in its NHTPC while retaining the same PCM fraction in the unit. This is a clear illustration of the importance of the design of the encapsulation structure in the overall heat transfer surface of the system.

Similarly, Nallusamy et al. [30] (Q) showed the highest NHTPC and PCM fraction combination calculated using HDPE spheres in a packed bed configuration.

4. Discussion

Data on the geometry, thermal properties, and performance under specific conditions of a wide range of technological approaches to LHES were gathered and used to estimate NHTPC and PCM volumetric fractions. Figure 6 summarizes the LHES units in each category to enable comparison of performance across all technologies.

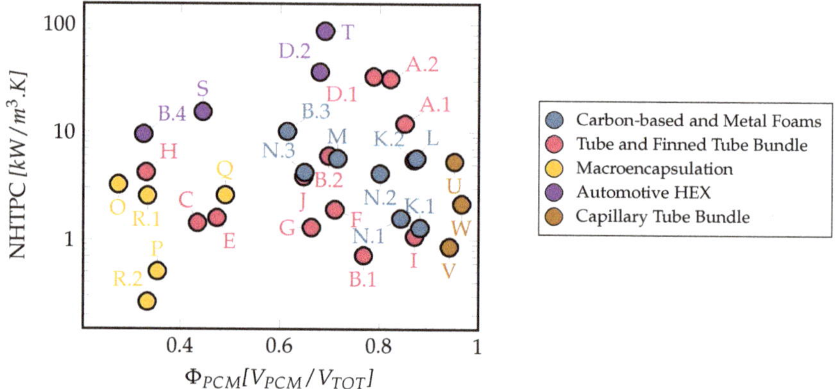

Figure 6. NHTPC vs. Φ_{PCM} for all units in each category in logarithmic scale.

Considering the overall trends when looking at the calculated KPIs, some general observations can be drawn:

- Finned tube bundles showed a wide range of performance, and volumetric fractions of PCM. The difference among the heat exchanger structures ranged mainly on terms of tube bundle geometry and fin density. If designed properly, they have a very high potential both in terms of heat transfer and resulting capacities.
- AHEX based units showed the highest average performance parameters across all technologies. This coupled with the advantages of mass production, potential for modularity, and variety of configurations makes them interesting for further study. Long-term stability and issues with flexibility of existing designs will have to be addressed before they are proposed for implementation in commercial products.
- Additionally, CTBs showed the potential to attain performances comparable to average performing finned tube bundle geometries while using a minimal share of the volume in a unit. However, the quality of the outlet temperature stability and round trip efficiency due to required pumping power at low diameters, as well as particularities regarding material compatibility and consequential limited temperature range, persist as some of the challenges for their implementation.
- Metallic foams and carbon-based TCE of all forms showed promising potential and further experimentation on optimized LHES unit structures, or simulations are required to properly assess their thermal performance.
- In contrast, among all the categories, the macroencapsulation techniques considered showed the lowest average NHTPC and Φ_{PCM} pairs. However, the studied units were not conceived or optimized for high power applications and used relatively large capsules with a generic geometry. A large room for improvement remains untapped regarding the optimization of capsule shape and size to enhance both heat transfer performance and compactness.

KPI Comparison

Considering the many already available KPI for LHES systems, Table 5 compiles some of the most relevant KPIs concerning heat transfer performance and compactness degree, as well as the pair presented in this study for three systems at different experimental conditions. Even though the considered KPIs are not intended to represent the same phenomena or were conceived with the same objectives, it is still interesting to see how they vary with experimental conditions and how they compare among each other.

Similarly to the robustness testing section, Units 2, 3, and 4 are used as representative systems, with varying inlet temperature, mass flow rate, and overall size, with the intent to analyze how the KPIs change accordingly.

Table 5. Comparison of different KPIs for LHES.

KPI	Unit 2 (T_{in}, 15 °C)	Unit 2 (T_{in}, 40 ° C)	Unit 3 ($\dot{m}$, 2088 kg/h)	Unit 3 ($\dot{m}$, 6984 kg/h)	Unit 4 ($\dot{m}$, 24 kg/h)	Unit 4 ($\dot{m}$, 62.4 kg/h)
Energy Stored [kWh]	2.10	1.70	18.77	18.77	0.23	0.24
Energy Density [kW/m^3·K]	105.16	85.23	47.93	47.93	26.19	27.23
NHTPC [kW/m^3·K]	31.39	31.81	1.60	1.56	4.22	4.36
Φ_{PCM}	0.83	0.83	0.66	0.66	0.33	0.33
t_{Disch} [h]	0.15	0.28	1.75	1.20	0.28	0.22
$\dot{Q}_{avg}$ [kW]	9.37	4.96	7.90	11.52	0.63	0.80
5min-Peak Power [kW]	16.37	8.67	23.25	35.69	0.86	1.51
5m-P.Power-Energy sto.[1/h]	7.78	5.09	1.24	1.90	3.80	6.40
γ [1/h]	4.46	2.91	0.42	0.61	2.79	3.40
ε_{avg}	0.77	0.78	0.41	0.13	0.73	0.58
h_v [kW/m^3·K]	16.85	20.70	3.68	16.48	9.17	14.54

As shown in the table, ED is intrinsically dependent not only on material properties and dimensions, but also on the sensible contributions and thus temperature levels imposed on the unit, as seen in the results for Unit 2 at two temperature levels. With a similar point of view, Φ_{PCM} shows a similar trend between the Units, but focusing only on the share of main energy storage material. Although it does not provide the exact same information, it can be useful in representing both the compactness degree of the design, and the potential for energy density of the unit without considering the current experimental temperature differences.

On the heat transfer performance side, specific charging rate γ as proposed by Guo et al. [39], and 5 min peak power-energy stored ratio by Gasia et al. [38] seem to agree with the overall trend shown by NHTPC across the units. They provide indications of the average heat transfer behavior with respect to the total energy, but remain dependent on both temperature and HTF mass flow rate differences. In contrast, NHTPC remains almost constant for a given system across different experimental conditions.

Although they all provide very useful information on particular aspects, slightly more drastic variations can be seen with the volumetric heat transfer coefficient h_v and average temperature effectiveness ε_{avg} presented by Nomura et al. [40] as well as t_{Disch}, $\dot{Q}_{avg}$ and 5 min. Peak Power.

5. Conclusions

A pair of performance indicators to evaluate the heat transfer performance and compactness for latent heat energy storage (LHES) units were presented. These key performance indicators (KPIs) were calculated for several LHES units reported in literature allowing a leveled performance comparison with regard to operating conditions at different scales, while remaining intrinsic to the geometry, heat exchanger structure materials, and PCM. The robustness of the proposed KPIs was confirmed with varying HTF mass flow rate and temperature levels, for units at different size scales, with a coefficient of variance below 6% in every case, and were compared to other reported KPIs for LHES.

Finned tube bundle (FTB) and tube bundle (TB) units showed the widest range of performance, and a great potential mainly dependent on the quality of the HEX design. Composites in general require further experimental work but show very promising potential.

Automotive heat exchanger (AHEX)-based units showed promise especially on their heat transfer performance, and are interesting for further study as they are already mass produced in a very wide range of variations. Capillary tube bundles (CTBs) show great potential especially in terms of compactness, but due to the added practical challenges require some optimization work for ideal operation. For both AHEX and CTBs, material limitations regarding compatibility and operation range are some of the main concerns, and should be thoroughly considered in further studies.

The macro encapsulated (ME)-based systems considered showed in general low performance and compactness, but a very large potential for improvement and flexibility, especially in terms of capsule shape and size optimization to customize their performance, for instance.

It is necessary to highlight that the publications taken into account had clear application-oriented objectives. This implies that achieving the highest possible heat transfer rate was not the focus during their conception, but only the required performance for a specific application under given conditions. It is possible to infer that optimized versions of the mentioned technologies would deliver considerably higher performance indicators. Therefore, the conclusions drawn from this analysis cannot be considered as final in any way regarding the technological approaches but more so as a look at the general potential of each approach.

Additionally, a key aspect of the thermal response is relatively overlooked by the analysis, as the methodology proposed only accounts for the stability of the outlet temperature indirectly, within the approximation of the average outlet temperature. Further dimensionless analysis is required to account for this effect.

Author Contributions: Conceptualization, S.M. and R.W.; Formal analysis, W.D.-D.; Funding acquisition, A.S.; Investigation, W.D.-D.; Methodology, W.D.-D.; Project administration, W.D.-D. and A.S.; Resources, A.S.; Supervision, A.S. and J.W.; Writing–original draft, W.D.-D.; Writing–review and editing, A.S., S.M., R.W. and J.W. All authors have read and agreed to the published version of the manuscript.

Funding: This research was funded by the Swiss Competence Center for Energy Research Heat and Electricity Storage (SCCER HaE).

Acknowledgments: This work was developed within the framework of the Swiss Competence Center for Energy Research Storage of Heat and Electricity (SCCER HaE).

Conflicts of Interest: The authors declare no conflict of interest. The funders had no role in the design of the study; in the collection, analyses, or interpretation of data; in the writing of the manuscript, or in the decision to publish the results.

Appendix A

Table A1. Additional information.

Type	T_{init}[°C]	$\overline{T_{HTF,In}}$ [°C]	$\overline{T_{HTF,Out}}$ [°C]	HTF	PCM	V_{PCM} [m^3]	HEX	Container	Ref.
TB (A.1)	68.0	15.0	33.85	Water	$CH_3COONa \cdot 3H_2O$	1.7×10^{-2}	Cu	PP	[7]
FTB (A.2)	68.0	16.0	48.65	Water	$CH_3COONa \cdot 3H_2O$	1.7×10^{-2}	Cu/Al	PP	[7]
TB (B.1)	55.0	20.0	20.02	Water	RT35	7.1×10^{-4}	Cu	Methacrylate	[48]
FTB (B.2)	55.0	20.0	20.18	Water	RT35	6.5×10^{-4}	Cu	Methacrylate	[48]
Triplex FTB(C)	92.0	68.0	69.09	Water	RT 82	6.8×10^{-3}	Cu	Cu	[9]
FTB (D.1)	45.0	100.0	99.72	Water	Xylitol	2.8×10^{-3}	Cu	No Info. Asm. Cu	[14]
FTB (E.1)	330.0	280.0	282.58	Thermal Oil	NaNO3	1.7×10^{-1}	Steel/Al alloy	No Info. Asm. Steel	[10]
FTB (F)	162.0	122.0	134.32	Hi-Tech Therm 60	KNO3/ NaNO/ NaNO2	6.2×10^{-2}	Cu	SS304	[8]
FTB (G)	155.0	105.0	113.25	Marlotherm SH	HDPE	2.6×10^{-1}	Steel / AlMg2.5	Steel	[47]
FTB (H)	72.0	20.0	44.12	Water	Organic	2.9×10^{-3}	Al	Glass	[13]
FTB (I)	60.0	10.0	18.71	Water	RT44HC	5.3×10^{-2}	Cu	Cu	[49]
FTB (J)	190.0	90.0	96.37	Hytherm 600	A164	5.4×10^{-3}	SS316	SS316	[12]
Graphite matrix (B.3)	55.0	20.0	20.18	Water	RT35	6.5×10^{-4}	Cu	Methacrylate	[48]
TB (K.1)	55.0	38.0	43.52	Water	Organic	1.8×10^{-2}	Cu	No Info. Asm. Cu	[55]
Carbon brushes (K.2)	55.0	38.0	48.43	Water	Organic	1.8×10^{-2}	Cu	No Info. Asm. Cu	[55]
Carbon cloth (L)	55.0	38.0	48.48	Water	Organic	1.8×10^{-2}	Cu	No Info. Asm. Cu	[19]
Graphite matrix (M)	65.0	25.0	50.96	Water	Paraffin	1.5×10^{-2}	Cu/Exp Graphite	None req.	[16]
TB (N.1)	70.0	25.0	36.51	Air	PCM 58P	5.2×10^{-4}	Al / Foam	Al	[22]
MF 95% porosity (N.2)	70.0	25.0	47.23	Air	PCM 58P	5.0×10^{-4}	Al / Foam	Al	[22]
MF 77% porosity (N.3)	70.0	25.0	47.18	Air	PCM 58P	4.0×10^{-4}	Al /Foam	Al	[22]

Table A1. *Cont.*

Type	T_{init}[°C]	$\overline{T_{HTF,In}}$ [°C]	$\overline{T_{HTF,Out}}$ [°C]	HTF	PCM	V_{PCM} [m^3]	HEX	Container	Ref.
ME (O)	627.0	527.0	571.85	Air	Al-25 wt%-Si	1.2×10^{-3}	AISI 316	AISI 316	[27]
ME (P)	535.0	380.0	416.06	Air	Na2SO4 -KCl	6.8×10^{-3}	Carbon steel	Carbon steel	[28]
ME (Q)	70.0	32.0	48.30	Water	Myristic Acid	2.3×10^{-2}	HDPE	Steel	[30]
Pouch ME (R.1)	75.0	45.0	51.28	Water	1-Octa -decanol	9.0×10^{-2}	Laminated PE, PA, Al, and PET	No Info. Asm. Steel	[31]
Sphere ME (R.2)	75.0	45.0	45.85	Water	1-Octa -decanol	9.0×10^{-2}	Laminated PE, PA, Al, and PET	No Info. Asm. Steel	[31]
AHEX (B.4)	55.0	20.0	21.29	Water	RT35	1.4×10^{-3}	Al/Cu	Methacrylate	[48]
AHEX (S)	3.5	24.0	17.08	Air	Organic	1.8×10^{-3}	Al	No Info. Asm. Al	[50]
AHEX (T)	70.0	28.3	41.87	Water	Stearic Acid	2.8×10^{-3}	Al	No Info. Asm. Al	[51]
AHEX (D.2)	45.0	100.0	99.83	90% Eth-Glycol	Xylitol	1.3×10^{-3}	AL1100	No Info. Asm. Al	[14]
CTB (U)	30.0	25.0	28.63	Water	Organic	9.7×10^{-1}	PP	No Info. Asm. PP	[52]
CTB (V)	33.0	22.0	24.58	Water	CaCl2 ·6H2O	8.0×10^{-1}	PP	PE	[53]
CTB (W)	35.0	22.0	25.75	Water	CaCl2 ·6H2O	1.4	PP	PVC	[54]

References

1. Child, M.; Bogdanov, D.; Breyer, C. The role of storage technologies for the transition to a 100% renewable energy system in Europe. *Energy Procedia* **2018**, *155*, 44–60, doi:10.1016/j.egypro.2018.11.067.
2. Prognos AG; Infras AG; TEP Energy GmbH; Bundesamt für Energie. *Analyse des Schweizerischen Energieverbrauchs 2000–2017*; Federal Office for Energy: Bern, Switzerland, 2018.
3. Heat Roadmap Europe. Heating and Cooling. Facts and Figures. The Transformation Towards a Low-carbon Heating & Cooling Sector. 2017. Available online: https://heatroadmap.eu/heating-and-cooling-energy-demand-profiles/ (accessed on 3 May 2020).
4. Benato, A.; Stoppato, A. Pumped Thermal Electricity Storage: A technology overview. *Therm. Sci. Eng. Prog.* **2018**, *6*, 301–315, doi:10.1016/j.tsep.2018.01.017.
5. Miró, L.; Gasia, J.; Cabeza, L.F. Thermal energy storage (TES) for industrial waste heat (IWH) recovery: A review. *Appl. Energy* **2016**, *179*, 284–301, doi:10.1016/j.apenergy.2016.06.147.
6. Seddegh, S.; Wang, X.; Henderson, A.D. Numerical investigation of heat transfer mechanism in a vertical shell and tube latent heat energy storage system. *Appl. Therm. Eng.* **2015**, *87*, 698–706, doi:10.1016/j.applthermaleng.2015.05.067.
7. Waser, R.; Maranda, S.; Stamatiou, A.; Zaglio, M.; Worlitschek, J. Modeling of solidification including supercooling effects in a fin-tube heat exchanger based latent heat storage. *Sol. Energy* **2018**, 1–12, doi:10.1016/j.solener.2018.12.020.
8. Niyas, H.; Rao, C.R.C.; Muthukumar, P. Performance investigation of a lab-scale latent heat storage prototype—Experimental results. *Sol. Energy* **2017**, *155*, 971–984, doi:10.1016/j.solener.2017.07.044.
9. Al-Abidi, A.A.; Mat, S.; Sopian, K.; Sulaiman, M.Y.; Mohammad, A.T. Experimental study of melting and solidification of PCM in a triplex tube heat exchanger with fins. *Energy Build.* **2014**, *68*, 33–41, doi:10.1016/j.enbuild.2013.09.007.
10. Laing, D.; Bauer, T.; Breidenbach, N.; Hachmann, B.; Johnson, M. Development of high temperature phase-change-material storages. *Appl. Energy* **2013**, *109*, 497–504, doi:10.1016/j.apenergy.2012.11.063.

11. Johnson, M.; Vogel, J.; Hempel, M.; Hachmann, B.; Dengel, A. Design of high temperature thermal energy storage for high power levels. *Sustain. Cities Soc.* **2017**, *35*, 758–763, doi:10.1016/j.scs.2017.09.007.
12. Raul, A.K.; Bhavsar, P.; Saha, S.K. Experimental study on discharging performance of vertical multitube shell and tube latent heat thermal energy storage. *J. Energy Storage* **2018**, *20*, 279–288, doi:10.1016/j.est.2018.09.022.
13. Amagour, M.E.H.; Rachek, A.; Bennajah, M.; Ebn Touhami, M. Experimental investigation and comparative performance analysis of a compact finned-tube heat exchanger uniformly filled with a phase change material for thermal energy storage. *Energy Convers. Manag.* **2018**, *165*, 137–151, doi:10.1016/j.enconman.2018.03.041.
14. Shon, J.; Kim, H.; Lee, K. Improved heat storage rate for an automobile coolant waste heat recovery system using phase-change material in a fin-tube heat exchanger. *Appl. Energy* **2014**, *113*, 680–689, doi:10.1016/j.apenergy.2013.07.049.
15. Xia, L.; Zhang, P.; Wang, R.Z. Preparation and thermal characterization of expanded graphite/paraffin composite phase change material. *Carbon* **2010**, *48*, 2538–2548, doi:10.1016/j.carbon.2010.03.030.
16. Wu, J.; Feng, Y.; Liu, C.; Li, H. Heat transfer characteristics of an expanded graphite / para ffi n PCM-heat exchanger used in an instantaneous heat pump water heater. *Appl. Therm. Eng.* **2018**, *142*, 644–655, doi:10.1016/j.applthermaleng.2018.06.087.
17. Nomura, T.; Tabuchi, K.; Zhu, C.; Sheng, N.; Wang, S.; Akiyama, T. High thermal conductivity phase change composite with percolating carbon fiber network. *Appl. Energy* **2015**, *154*, 678–685, doi:10.1016/j.apenergy.2015.05.042.
18. Karthik, M.; Faik, A.; Blanco-Rodríguez, P.; Rodríguez-Aseguinolaza, J.; D'Aguanno, B. Preparation of erythritol–graphite foam phase change composite with enhanced thermal conductivity for thermal energy storage applications. *Carbon* **2015**, *94*, 266–276, doi:10.1016/j.carbon.2015.06.075.
19. Nakaso, K.; Teshima, H.; Yoshimura, A.; Nogami, S.; Hamada, Y.; Fukai, J. Extension of heat transfer area using carbon fiber cloths in latent heat thermal energy storage tanks. *Chem. Eng. Process. Process Intensif.* **2008**, *47*, 879–885, doi:10.1016/j.cep.2007.02.001.
20. Xiao, M.; Feng, B.; Gong, K. Thermal performance of a high conductive shape-stabilized thermal storage material. *Sol. Energy Mater. Sol. Cells* **2001**, *69*, 293–296, doi:10.1016/S0927-0248(01)00056-3.
21. Lazzarin, R.M.; Mancin, S.; Noro, M.; Righetti, G. Hybrid PCM-aluminium foams' thermal storages: An experimental study. *Int. J. Low Carbon Technol.* **2018**, *13*, 286–291, doi:10.1093/IJLCT/CTY030.
22. Atal, A.; Wang, Y.; Harsha, M.; Sengupta, S. Effect of porosity of conducting matrix on a phase change energy storage device. *Int. J. Heat Mass Transf.* **2016**, *93*, 9–16, doi:10.1016/j.ijheatmasstransfer.2015.09.033.
23. Mancin, S.; Diani, A.; Doretti, L.; Hooman, K.; Rossetto, L. Experimental analysis of phase change phenomenon of paraffin waxes embedded in copper foams. *Int. J. Therm. Sci.* **2015**, *90*, 79–89, doi:10.1016/j.ijthermalsci.2014.11.023.
24. Xiao, X.; Zhang, P.; Li, M. Effective thermal conductivity of open-cell metal foams impregnated with pure paraffin for latent heat storage. *Int. J. Therm. Sci.* **2014**, *81*, 94–105, doi:10.1016/j.ijthermalsci.2014.03.006.
25. Thapa, S.; Chukwu, S.; Khaliq, A.; Weiss, L. Fabrication and analysis of small-scale thermal energy storage with conductivity enhancement. *Energy Convers. Manag.* **2014**, *79*, 161–170, doi:10.1016/j.enconman.2013.12.019.
26. Oya, T.; Nomura, T.; Okinaka, N.; Akiyama, T. Phase change composite based on porous nickel and erythritol. *Appl. Therm. Eng.* **2012**, *40*, 373–377, doi:10.1016/j.applthermaleng.2012.02.033.
27. Ma, F.; Zhang, P. Investigation on the performance of a high-temperature packed bed latent heat thermal energy storage system using Al-Si alloy. *Energy Convers. Manag.* **2017**, *150*, 500–514, doi:10.1016/j.enconman.2017.08.040.
28. Wickramaratne, C.; Moloney, F.; Pirasaci, T.; Kamal, R.; Goswami, D.; Stefanakos, E.; Dhau, J. Experimental Study on Thermal Storage Performance of Cylindrically Encapsulated PCM in a Cylindrical Storage Tank With Axial Flow. In Proceedings of the ASME 2016 Power Conference Collocated with the ASME 2016 10th International Conference on Energy Sustainability and the ASME 2016 14th International Conference on Fuel Cell Science, Engineering and Technology, Charlotte, NC, USA, 26–30 June 2016; p. V001T08A014, doi:10.1115/POWER2016-59427.
29. Pirasaci, T.; Wickramaratne, C.; Moloney, F.; Goswami, D.Y.; Stefanakos, E. Influence of design on performance of a latent heat storage system for a direct steam generation power plant. *Appl. Energy* **2018**, *224*, 220–229, doi:10.1016/j.apenergy.2018.04.122.

30. Nallusamy, N.; Sampath, S.; Velraj, R. Experimental investigation on a combined sensible and latent heat storage system integrated with constant/varying (solar) heat sources. *Renew. Energy* **2007**, *32*, 1206–1227, doi:10.1016/j.renene.2006.04.015.
31. Park, J.; Shin, D.H.; Lee, S.J.; Shin, Y.; Karng, S.W. Effective latent heat thermal energy storage system using thin flexible pouches. *Sustain. Cities Soc.* **2018**, doi:10.1016/j.scs.2018.10.046.
32. Jesumathy, S.; Udayakumar, M.; Suresh, S. Experimental study of enhanced heat transfer by addition of CuO nanoparticle. *Heat Mass Transf. Und Stoffuebertragung* **2012**, *48*, 965–978, doi:10.1007/s00231-011-0945-y.
33. Harikrishnan, S.; Magesh, S.; Kalaiselvam, S. Preparation and thermal energy storage behaviour of stearic acid—TiO_2 nanofluids as a phase change material for solar heating systems. *Thermochim. Acta* **2013**, *565*, 137–145, doi:10.1016/j.tca.2013.05.001.
34. Harikrishnan, S.; Imran Hussain, S.; Devaraju, A.; Sivasamy, P.; Kalaiselvam, S. Improved performance of a newly prepared nano-enhanced phase change material for solar energy storage. *J. Mech. Sci. Technol.* **2017**, *31*, 4903–4910, doi:10.1007/s12206-017-0938-y.
35. Romaní, J.; Gasia, J.; Solé, A.; Takasu, H.; Kato, Y.; Cabeza, L.F. Evaluation of energy density as performance indicator for thermal energy storage at material and system levels. *Appl. Energy* **2019**, *235*, 954–962, doi:10.1016/j.apenergy.2018.11.029.
36. Wang, W.W.; Wang, L.B.; He, Y.L. The energy efficiency ratio of heat storage in one shell-and-one tube phase change thermal energy storage unit. *Appl. Energy* **2015**, *138*, 169–182, doi:10.1016/j.apenergy.2014.10.064.
37. Li, G. Energy and exergy performance assessments for latent heat thermal energy storage systems. *Renew. Sustain. Energy Rev.* **2015**, *51*, 926–954, doi:10.1016/j.rser.2015.06.052.
38. Gasia, J.; Diriken, J.; Bourke, M.; Van Bael, J.; Cabeza, L.F. Comparative study of the thermal performance of four different shell-and-tube heat exchangers used as latent heat thermal energy storage systems. *Renew. Energy* **2017**, *114*, 934–944, doi:10.1016/j.renene.2017.07.114.
39. Guo, X.; Goumba, A.P. Process intensification principles applied to thermal energy storage systems-A brief review. *Front. Energy Res.* **2018**, *6*, 1–10, doi:10.3389/fenrg.2018.00017.
40. Nomura, T.; Tsubota, M.; Oya, T.; Okinaka, N.; Akiyama, T. Heat release performance of direct-contact heat exchanger with erythritol as phase change material. *Appl. Therm. Eng.* **2013**, *61*, 28–35, doi:10.1016/j.applthermaleng.2013.07.024.
41. Krimmel, S.; Stamatiou, A.; Worlitschek, J.; Walter, H. Experimental characterization of the heat transfer in a latent direct contact thermal energy storage with one nozzle in labor scale. *Int. J. Mech. Eng.* **2018**, *3*, 83–97.
42. Incropera, F.P. *Fundamentals of Heat and Mass Transfer*; John Wiley and Sons, Inc.: Hoboken, NJ, USA, 2006.
43. Tay, N.H.; Belusko, M.; Bruno, F. An effectiveness-NTU technique for characterising tube-in-tank phase change thermal energy storage systems. *Appl. Energy* **2012**, *91*, 309–319, doi:10.1016/j.apenergy.2011.09.039.
44. Tay, N.H.; Belusko, M.; Castell, A.; Cabeza, L.F.; Bruno, F. An effectiveness-NTU technique for characterising a finned tubes PCM system using a CFD model. *Appl. Energy* **2014**, *131*, 377–385, doi:10.1016/j.apenergy.2014.06.041.
45. Incropera, F.; DeWitt, D.; Bergman, T.; Lavine, A. *Fundamentals of Heat and Mass Transfer*; Wiley: Hoboken, NJ, USA, 2007.
46. Mao, Q.; Liu, N.; Peng, L. Numerical Investigations on Charging / Discharging Performance of a Novel Truncated Cone Thermal Energy Storage Tank on a Concentrated Solar Power System *Int. J. Photoenergy* **2019**. *2019*.
47. Zauner, C.; Hengstberger, F.; Etzel, M.; Lager, D.; Hofmann, R.; Walter, H. Experimental characterization and simulation of a fin-tube latent heat storage using high density polyethylene as PCM. *Appl. Energy* **2016**, *179*, 237–246, doi:10.1016/j.apenergy.2016.06.138.
48. Medrano, M.; Yilmaz, M.O.; Nogués, M.; Martorell, I.; Roca, J.; Cabeza, L.F. Experimental evaluation of commercial heat exchangers for use as PCM thermal storage systems. *Appl. Energy* **2009**, *86*, 2047–2055, doi:10.1016/j.apenergy.2009.01.014.
49. Khan, Z.; Khan, Z.A. An experimental investigation of discharge/solidification cycle of paraffin in novel shell and tube with longitudinal fins based latent heat storage system. *Energy Convers. Manag.* **2017**, *154*, 157–167, doi:10.1016/j.enconman.2017.10.051.
50. Lee, D.; Kim, J.; Yim, E.; Jeon, C.; Jung, C.; Han, B. Experimental study on performance characteristics of cold storage heat exchanger for ISG vehicle. *Int. J.* **2012**, *13*, 293–300, doi:10.1007/s12239.

51. Park, S.; Woo, S.; Shon, J.; Lee, K. Experimental study on heat storage system using phase-change material in a diesel engine. *Energy* **2017**, *119*, 1108–1118, doi:10.1016/j.energy.2016.11.063.
52. Hejcik, J.; Charvat, P.; Klimes, L.; Astrouski, I. A Pcm-Water Heat Exchanger with Polymeric Hollow Fibres for Latent Heat Thermal Energy Storage: A Parametric Study of Discharging Stage. *J. Theor. Appl. Mech.* **2016**, *54*, 1285–1295, doi:10.15632/jtam-pl.54.4.1285.
53. Helm, M.; Keil, C.; Hiebler, S.; Mehling, H.; Schweigler, C. Solar heating and cooling system with absorption chiller and low temperature latent heat storage: Energetic performance and operational experience. *Int. J. Refrig.* **2009**, *32*, 596–606, doi:10.1016/j.ijrefrig.2009.02.010.
54. Helm, M.; Hagel, K.; Pfeffer, W.; Hiebler, S.; Schweigler, C. Solar heating and cooling system with absorption chiller and latent heat storage—A research project summary. *Energy Procedia* **2014**, *48*, 837–849, doi:10.1016/j.egypro.2014.02.097.
55. Fukai, J.; Hamada, Y.; Morozumi, Y.; Miyatake, O. Improvement of thermal characteristics of latent heat thermal energy storage units using carbon-fiber brushes: Experiments and modeling. *Int. J. Heat Mass Transf.* **2003**, *46*, 4513–4525, doi:10.1016/S0017-9310(03)00290-4.

Article

Selection of the Appropriate Phase Change Material for Two Innovative Compact Energy Storage Systems in Residential Buildings

Gabriel Zsembinszki, Angel G. Fernández and Luisa F. Cabeza *

GREiA Research Group, Universitat de Lleida, Pere de Cabrera s/n, 25001-Lleida, Spain; gabriel.zsembinszki@udl.cat (G.Z.); angel.fernandez@udl.cat (A.G.F.)

* Correspondence: luisaf.cabeza@udl.cat; Tel.: +34-973-003576

Received: 13 February 2020; Accepted: 17 March 2020; Published: 20 March 2020

Featured Application: The selection of suitable phase change material as thermal energy storage for space cooling, heating, or domestic hot water in buildings.

Abstract: The implementation of thermal energy storage systems using phase change materials to support the integration of renewable energies is a key element that allows reducing the energy consumption in buildings by increasing self-consumption and system efficiency. The selection of the most suitable phase change material is an important part of the successful implementation of the thermal energy storage system. The aim of this paper is to present the methodology used to assess the suitability of potential phase change materials to be used in two innovative energy storage systems, one of them being mainly intended to provide cooling, while the other provides heating and domestic hot water to residential buildings. The selection methodology relies on a qualitative decision matrix, which uses some common features of phase change materials to assign an overall score to each material that should allow comparing the different options. Experimental characterization of the best candidates was also performed to help in making a final decision. The results indicate some of the most suitable candidates for both systems, with RT4 being the most promising commercial phase change material for the system designed to provide cooling, while for the system designed to provide heating and domestic hot water, the most promising candidate is RT64HC, another commercial product.

Keywords: thermal energy storage (TES); phase change material (PCM); heating and cooling; material selection; selection methodology

---

1. Introduction

The use of thermal energy storage (TES) systems in applications in which either energy supply is intermittent or energy demand has large fluctuations is a key aspect of any energy system, able to reduce energy consumption or costs, besides enhancing the flexibility of the system and reducing the dependence on the energy source availability [1]. The importance of the implementation of TES systems to support the integration of renewable energies in buildings was made evident by different initiatives of the European Commission aimed at achieving a considerable reduction of energy consumption in buildings by increasing self-consumption [2].

TES technologies have been developed for applications that cover a wide temperature range, from low-temperature applications such as ice storage or the conservation and transport of temperature-sensitive materials [3,4], going through medium temperature applications such as space heating and cooling in buildings [5,6], and domestic hot water (DHW) generation [7–9], up to high-temperature applications such as concentrated solar power (CSP) plants [10,11].

Among the different TES technologies available or under investigation, latent heat storage using solid–liquid phase change materials (PCM) has some advantages over sensible heat storage because it has higher heat storage capacity and a more stable thermal behavior during both charging and discharging processes. There are many studies available in the literature regarding the main aspects of the use of PCM as TES material in different applications [12–14].

Since the PCM selection procedure is an important step to achieve proper and reliable operation of the TES system, further efforts are needed to ensure that this step is done correctly. For this purpose, most studies rely on the main PCM thermophysical properties such as melting temperature, enthalpy, density, and conductivity. However, other properties such as health hazards, thermal cycling, and stability should also be taken into account to complement the most common selection criteria [15,16]. The PCM selection methodology proposed by Miró et al. [15] and Gasia et al. [16] requires a comprehensive study of different properties and aspects regarding the potential PCM candidates, which should include a detailed experimental characterization. Nevertheless, when the number of PCM candidates is quite large, a preliminary selection based on information provided by the PCM manufacturers or available in the literature is deemed necessary before conducting a more thorough experimental analysis of the most suitable PCM candidates.

The objective of this paper is to present the methodology and the results obtained for the initial step of an overall evaluation and materials selection process of the most suitable PCM to be used in two innovative systems that implement a compact hybrid electrical and thermal energy storage, intended to provide heating, cooling, and DHW in residential buildings in cold and warm weather conditions.

2. Methodology

2.1. Description of the Application

The latent heat TES investigated in this study is part of an innovative concept of two hybrid electrical and thermal energy storage systems developed and tested in the framework of HYBUILD, a research project funded by the H2020 program [17]. The main aim of the systems is to provide heating or cooling to stand-alone and district connected residential buildings by using a considerable contribution from solar energy to reduce the energy consumption of the buildings [18]. One of the concepts is mainly intended to cover the cooling demand in buildings located in Mediterranean climate regions, while the other concept is mainly designed to provide heating and DHW for buildings located in Continental climate regions. Both systems contain an electrical and thermal energy storage system to increase the use of solar energy.

In the Mediterranean concept (MED), a latent heat TES using PCM is located in the low-pressure side of a compression heat pump that is connected, by means of a DC bus, to a field of PV panels. In this way, the surplus of cold produced by the heat pump during periods of high solar radiation availability can be stored inside the TES tank, and it can be used in subsequent periods when no or insufficient solar energy is available. Since the latent heat TES is connected to both the building cooling loop and the chiller, the ideal phase change temperature of the PCM should be around 4 °C to allow a sufficient temperature difference between the PCM and the cooling water, and also an efficient charging of the PCM by the refrigerant of the chiller that evaporates inside the TES tank.

In the Continental concept (CON), the latent heat TES is located at the compressor discharge of a compression heat pump to store the sensible heat of the hot refrigerant gas before it condenses inside the condenser. Similar to the MED concept, in this case, the compression heat pump is also connected to a field of PV panels by means of a DC bus. The TES tank can be charged during both heating and cooling operation modes, and the heat stored in the PCM is used to heat the DHW up to the desired temperature. Since, in this case, the latent heat TES is used to support DHW generation, a PCM phase change temperature around 65 °C is expected in this case.

2.2. PCM Selection Methodology

The first step of the selection methodology consisted of a thorough review of potential PCM candidates available in the literature, whether commercial or not, for both MED and CON systems. Therefore, the initial search was performed focusing only on the phase change temperature of the PCM. A wider temperature range, from 0 °C to 7 °C and from 48 °C to 68 °C, was initially considered for the MED and CON systems, respectively, due to some uncertainties regarding the optimal phase change temperature at the very preliminary stage of the project development, when the PCM selection process began.

As a second step of the selection methodology, given the relatively large number of possible PCM candidates found in the first step, a pre-selection was performed in order to exclude PCMs that were not suitable either because of health hazard issues, corrosion with the tank container material (aluminum), or very bad thermophysical properties. Likewise, when two or more different PCM had similar features, only the most suitable of them was chosen.

Once a reduced group of potential PCM candidates was obtained, the third step of the methodology consisted of the development of a decision matrix as a tool to make a final selection based on objective criteria. Since there is no standard selection procedure, different selection criteria can be used, depending on the focus of the application. General properties are usually taken into account, such as phase change enthalpy, width of the phase change range, subcooling, hysteresis, thermal conductivity, density, availability, toxicity, or cost. Some of these properties were used in the pre-selection step to exclude some of the potential PCM candidates, therefore the decision matrix was built to compare the different valid options and to find the most promising ones for the two HYBUILD systems.

Hence, the following parameters were considered in the decision matrix: width of the phase change range, melting enthalpy, availability, price, and maximum working temperature in the case of the CON system. The selection of these parameters is justified by their direct impact on the system viability from the operational and economic points of view, and the relative easiness in obtaining their values. A score was assigned to each of the above decision parameters for each PCM candidate taking into account the criteria shown in Table 1.

Table 1. The scoring criteria applied to each decision parameter.

Temperature Range (T, in °C)		Enthalpy (h, in kJ/kg)		Availability (-)		Price (P, in €/kg)		Maximum Temperature (T_{max}, in °C)	
T < 2	3	h > 250	3	Yes	3	P < 2.5	3	T_{max} > 120	3
2 < T <3	2	200 < h < 250	2	No	0	2.5 < P < 5	2	T_{max} < 120 or n.a.	0
3 < T <4	1	150 < h < 200	1	-	-	5 < P < 10	1	-	-
T > 4 or n.a.	0	h < 150 or n.a.	0	-	-	P > 10 or n.a.	0	-	-

n.a.—not available.

Next, a total score was calculated for each of the PCM, based on a weighted average of the partial scores obtained for each of the decision parameters. With regards to the values chosen for the weight, two scenarios were considered, as shown in Table 2: Scenario 1, in which the value of each decision parameter was selected based on authors criteria based on their experience, and Scenario 2, in which all decision parameters were given the same weight. The weighting of the parameters in Scenario 1 is the one that suits the main requirements of the systems, such as compactness and cost-effectiveness of the TES solution, which make it so PCM enthalpy and price have a higher weight with respect to the other decision parameters.

Table 2. Reference values of the weights.

Decision Parameter	Weight (%)			
	MED System		CON System	
	Scenario 1	Scenario 2	Scenario 1	Scenario 2
Phase change range (C_p-T curve)	25	25	20	20
Enthalpy	30	25	25	20
Availability	15	25	10	20
Price	30	25	25	20
Maximum working temperature	-	-	20	20
Total	100	100	100	100

However, a sensitivity analysis was deemed necessary to check the influence that the variation of the different weights could have on the final result. This consisted of altering the reference values (Ref) of the weights shown in Table 2 by adding or subtracting a certain amount to the reference value of two or more decision parameters. All possible combinations were considered, which satisfied the condition that the sum of the absolute values of the deviations with respect to the reference values is equal to 20%. A total of 49 and 180 combinations of altered values were obtained for the MED and CON systems, respectively.

For each of the combinations obtained, the total score was calculated for all PCM candidates for both MED and CON systems, and the best three PCM of the ranking in each scenario was recorded. The average score obtained by each PCM candidate and the overall statistics for the whole set of combinations were performed to determine the significance of the results obtained using the decision matrix.

2.3. Thermophysical Characterization

The last step of the methodology was the analysis of the thermophysical properties of the best candidates to select the most suitable PCM for each system. The thermophysical characterization included the analysis with a differential scanning calorimeter DSC 3+ (Mettler Toledo, Schwerzenbach, Switzerland). The amount of sample used was around 10 mg and experiments were performed under N_2 flow. The PCM samples were located into 100 μL cold-welded aluminum crucibles. The methodology followed to obtain the phase change temperature and enthalpy of the PCM is based on dynamic temperature programs from 20 °C under their theoretical phase change temperatures to 20 °C above it. The equipment precision is ±0.1 °C for temperature and ± 3 J/g for enthalpy results.

Thermogravimetric analysis (TGA) was used to characterize the PCM maximum working temperature (defined as the temperature at which the material has lost 1.5 wt % of its mass) and the final degradation temperature. The equipment used was a TGA 2 (Mettler Toledo, Schwerzenbach, Switzerland), which allows measuring samples up to 1000 °C and has a sensitivity of 0.1 μg. The analyses were performed under a 50 mL/min N_2 atmosphere. The heating rate used to perform the thermogravimetric analysis was 10 °C/min from 25 °C to 100 °C (low-temperature PCM) and from 25 °C to 200 °C (high-temperature PCM). Opened 100 μL alumina crucibles used were filled with around 1/3 volume of material leading to average sample masses of around 10 mg.

The cycling stability of the selected PCMs was characterized using a thermal cycler GeneQ BIOER TC-18/H(b) (BIOER, Hangzhou, Zhejiang, China). Due to the operating temperature range of the thermal cycler, it was only possible to cycle the pre-selected PCM for the CON system. The cycles were carried out from 20 °C to 80 °C using a heating rate of 12.5 °C/min. Samples were taken after 10, 100, 1000, and 8000 cycles.

The chemical characterization of the PCMs before and after cycling was carried out using a Fourier transform infrared (FT–IR) spectroscopy with attenuated total reflectance (ATR), which analyses the PCM chemical degradation caused by thermal cycling. The advantage of ATR is the possibility of obtaining the spectra directly from the sample, without any specific sample preparation. The partial or total disappearance of the characteristic peaks and/or the appearance of new peaks can indicate that the material is being oxidized or degraded. This analysis was carried out with a PIKE MIRacle™

ATR (PIKE Technologies, Madison, WI, USA) sampling accessory with a Diamond/ZnSe ATR base, FT–IR 6300 (Hachioji, Tokyo, Japan). It allows analyzing substances in solid and liquid states. It was optimized by a wavelength range between 4000 cm^{-1} and 650 cm^{-1}, and its standard spectral resolution is 4 cm^{-1} accounting for 64 infrared scans for each analysis; the data recorded are their means. Its functionality is based on the characteristic wave numbers at which the molecules vibrate in infrared frequencies. Since no thermal cycling tests were performed with the PCM samples for the MED system, the FT–IR analysis was only done for the PCM candidates for the CON system.

3. Results and Discussion

3.1. Mediterranean System (Low Temperature)

For the MED system, the number of potential PCM candidates found in the literature review having a phase change temperature in the range between 0 °C and 7 °C was around 60. After applying some excluding criteria as explained in the methodology, this number was further reduced by half. Finally, the decision matrix was only applied to PCM with phase change temperature within the core temperature interval between 2 °C and 4 °C shown in Table 3.

Table 3. Pre-selected available phase change material (PCM) for the Mediterranean system and some of their thermophysical properties as reported in the literature.

Commercial Name/Composition	Type	Melting Temperature (°C)	Phase Change Enthalpy (kJ/kg)	Thermal Conductivity (W/m·K)	Density (kg/m^3)	Reference
RT3HC_1	Organic (paraffin)	1–3	190	0.20 (l) 0.20 (s)	770 (l) 880 (s)	[19]
A3	Organic (n.a.)	3	200	0.210	765	[20]
0200- Q2 BioPCM	Organic (bioPCM)	2	200–230	0.2–0.7 (l) 0.25–2.5 (s)	850–1300 (l) 900–1250 (s)	[21]
PCM-PDR03P	Organic (n.a.)	3.5	185	n.a.	570	[22]
savE OM 03	Organic (n.a.)	3.5	229	0.224 (l) 0.146 (l)	835 (l) 912 (s)	[23]
Caprylic acid + lauric acid (9:1 by mol)	Organic eutectic (fatty acid)	3.8	151.5	n.a.	n.a.	[24]
RT4	Organic (paraffin)	2–4	175	0.20 (l) 0.20 (s)	770 (l) 880 (s)	[19]
0200- Q4 BioPCM	Organic (bioPCM)	4	200–230	0.2–0.7 (l) 0.25–2.5 (s)	850–1300 (l) 900–1250 (s)	[21]
PureTemp 4	Organic (bio-based)	4	195	n.a.	n.a.	[25]
Tetrahydrofuran clathrate hydrate	Inorganic (clathrate hydrate)	4.4	255	n.a.	n.a.	[26]

n.a.—not available.

Figure 1 shows the results obtained by applying the decision matrix based on the scoring criteria shown in Table 1 and using the weights that correspond to Scenario 1 in Table 2. The best score was obtained by savE OM 03, a commercial product manufactured by the Indian company PLUSS® [23]. The second-best score was obtained by another commercial PCM, RT4, manufactured by the German company Rubitherm [19]. Three other commercial PCM, RT3HC_1 [19], A3 [20], and PureTemp 4 [25], share the third place of the ranking with very similar scores.

Furthermore, three of the pre-selected PCM candidates among those that obtained the highest score were characterized in the laboratory. Figure 2 shows the results of the DSC analysis of savE OM 03, RT4, and RT3HC_1 pre-selected PCM candidates.

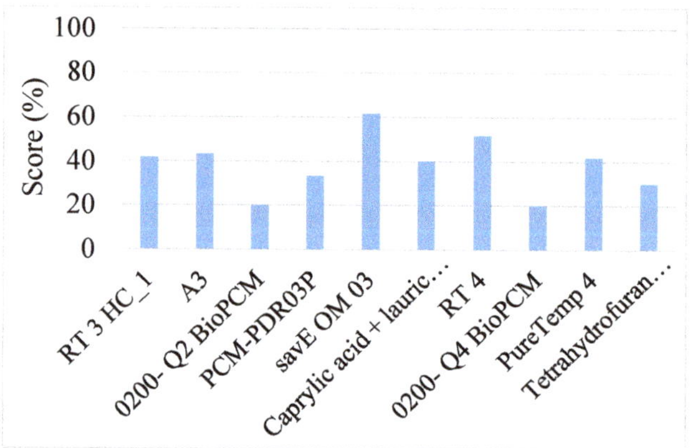

Figure 1. Score obtained by each of the pre-selected PCM candidates in Scenario 1 for the Mediterranean concept (MED) system.

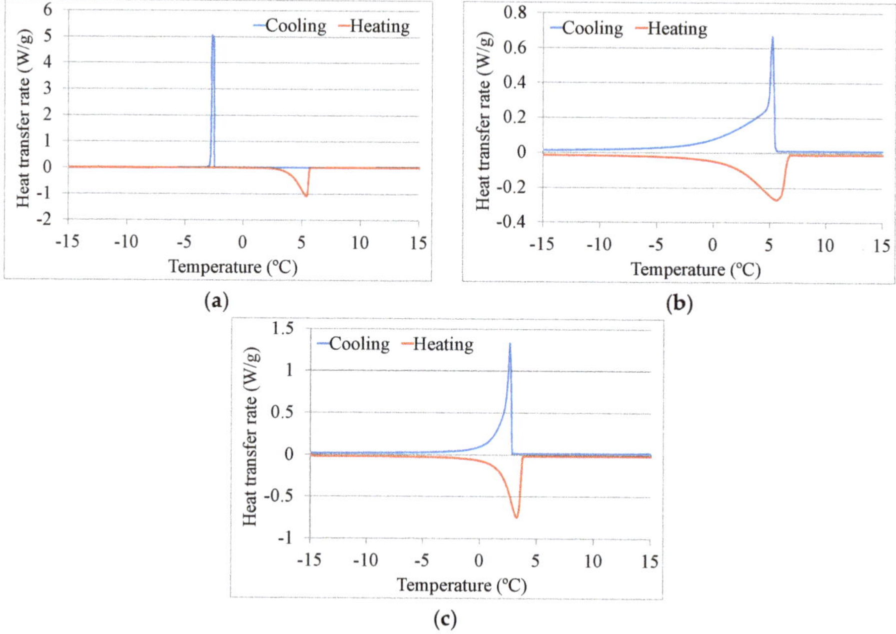

Figure 2. DSC analysis of three of the most promising pre-selected PCM for the MED system. (**a**) savE OM 03; (**b**) RT4; (**c**) RT3HC_1.

On the one hand, it can be seen that both PCM from Rubitherm have an acceptable behavior in both heating and cooling processes, with the temperature difference between the heating and cooling peaks being around 0.5 °C. Moreover, RT3HC_1 shows a sharper curve profile around the phase change temperature and a slightly higher phase change enthalpy than RT4. On the other hand, savE OM 03 has a heating curve more similar to the one for RT3HC_1, while the cooling curve is by far much sharper than the ones for the two PCM from Rubitherm. However, the difference between the heating and cooling peaks corresponding to melting and solidification curves for savE OM 03 is about 8 °C, which is a serious drawback for the implementation of this PCM in practical applications.

The results of the TGA performed for the same three PCM candidates for the MED system are shown in Figure 3. The mass loss of the three PCM is less than 0.1% when the temperature is below 48 °C and only RT3HC_1 reaches a mass loss of 1.5 wt % at 87 °C. At 100 °C. The mass losses for savE OM 03, RT4, and RT3HC_1 are 0.45 wt %, 0.91 wt %, and 2.15 wt %, respectively. If a lower heating rate of e.g., 1 °C/min was used instead of 10 °C/min, the temperature for the stability limit would be lower than the values shown here, but still, the mass loss in the temperature range of the application would be negligible. Therefore, from this point of view, the three PCM candidates are suitable because the ambient temperature is expected to be below 50 °C in practically all locations in the world.

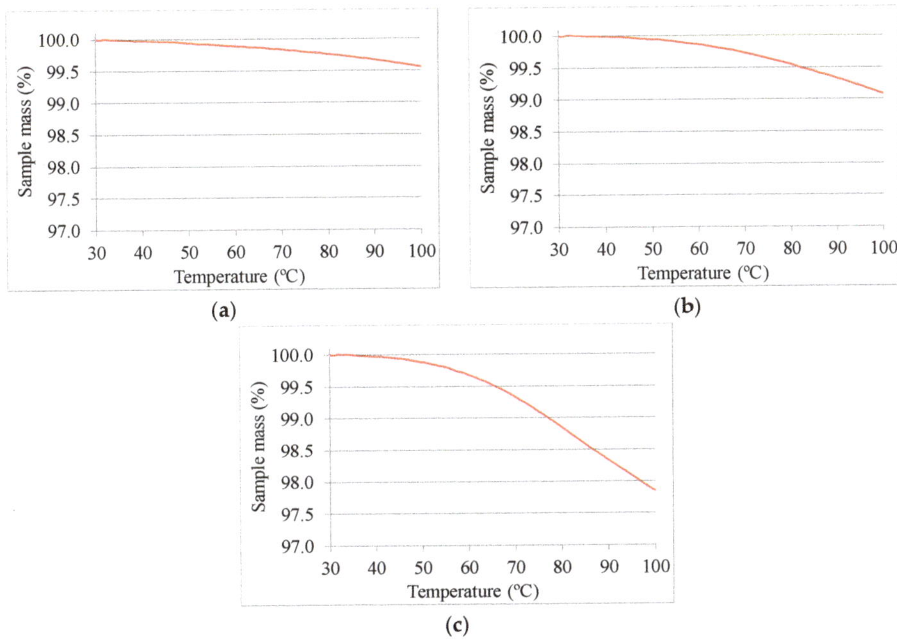

Figure 3. TGA of the PCM candidates for the MED system. (**a**) savE OM 03; (**b**) RT4; (**c**) RT3HC_1.

To summarize, the results given by the decision matrix indicate that savE OM 03, RT4, A3, RT3HC_1, and PureTemp 4 are the most promising options. Nevertheless, because of the very large hysteresis of around 8 °C, savE OM 03 is not a recommended option for the MED solution.

3.2. Continental System

As a result of the literature review, more than 120 PCM potential candidates were initially found for the CON system, which doubles the number found for the MED system. After the pre-selection procedure, the number of candidates was considerably reduced, so that only 25 PCM were taken into consideration in the decision matrix. The list of all pre-selected PCM candidates for the Continental system is shown in Table 4.

Table 4. Pre-selected PCM for the Continental system and some of their thermophysical properties as reported in the literature.

Commercial Name/Composition	Type	Melting Temperature (°C.)	Phase Change Enthalpy (kJ/kg)	Thermal Conductivity (W/m·K)	Density (kg/m^3)	Reference
A50	Organic (n.a.)	50	218	0.18	810	[20]
0500- Q50 BioPCM	Organic (bioPCM)	50	200–230	0.2–0.7 (l) 0.25–2.5 (s)	850–1300 (l) 900–1250 (s)	[21]
savE OM50	Organic (fatty acids mixture)	50–51	223	0.14 (l) 0.21 (s)	859 (l) 961 (s)	[23]
RT54HC	Organic (paraffin)	53–54	200	0.2	800 (l) 850 (s)	[19]
Stearic acid ($CH_3(CH_2)_{16}$-COOH)	Organic (fatty acid)	54	157	0.17 (l) 0.29 (s)	940 (s)	[27]
Cetyl stearate	Organic (ester)	54.6	212.1–216.3	n.a.	n.a.	[26]
savE OM 55	Organic (mixture of fatty acids)	55	208	0.16 (l) 0.1 (s)	841 (l) 935 (s)	[23]
0500- Q56 BioPCM	Organic (bioPCM)	56	200–230	0.2–0.7 (l) 0.25–2.5 (s)	850–1300 (l) 900–1250 (s)	[21]
Tristearin ($(C_{17}H_{35}COO)_3C_3H_5$)	Organic	56	190.8	n.a.	862 (l)	[26]
PureTemp 58	Organic (bio-based)	58	225	0.15 (l) 0.25 (s)	810 (l) 890 (s)	[25]
A58H	Organic (n.a.)	58	243	0.18	820	[20]
66.7% Polyethylene oxide 10000 + 33.3% Myristic acid	Organic (plastic + fatty acid)	58.7	191	n.a.	n.a.	[28]
Climsel C58	Inorganic (salt hydrate)	58	259	1.46	n.a.	[12,14]
		55–58	260	0.47 (l) 0.57 (s)	1400	[29]
		58	80	0.5–0.7	1460	[30]
THP 5860	Organic (paraffin)	55–60	153	n.a.	n.a.	Own measurements
Paraffin C27	Organic (paraffin)	58.8	236	n.a.	n.a.	[31]
RT60	Organic (paraffin)	58–60	214	0.2	n.a.	[30]
Stearyl stearate	Organic (ester)	59.2	214.75–214.93	n.a.	n.a.	[26]
PureTemp 63	Organic (bio-based)	63	206	0.15 (l) 0.25 (s)	840 (l) 920 (s)	[25]
RT64HC	Organic (n.a.)	63–65	250	0.2	780 (l) 880 (s)	[19]
Stearyl arachidate ($C_{38}H_{76}O_2$)	Organic (ester)	64.96	226	n.a.	2350 (l) 1930 (s)	[26]
50% CH_3CONH_2 + 50% $C_{17}H_{35}COOH$	Organic (eutectic)	65	218	n.a.	n.a.	[31]
0500- Q65 BioPCM	Organic (bioPCM)	65	200–230	0.2–0.7 (l) 0.25–2.5 (s)	850–1300 (l) 900–1250 (s)	[21]
savE FS 65	Organic (blend of organic material in polymer matrix)	66–68	218	0.25 (s)	842 (s)	[23]
PureTemp 68	Organic (bio-based)	68	213	0.15 (l) 0.25 (s)	870 (l) 960 (s)	[25]
0500- Q68 BioPCM	Organic (bioPCM)	68	200-235	0.2–0.7 (l) 0.25–2.5 (s)	850–1300 (l) 900–1250 (s)	[21]

Figure 4 shows the results obtained after applying the decision matrix based on the scoring criteria shown in Table 1 and using the weights that correspond to Scenario 1 in Table 2 only to pre-selected PCM with a phase change temperature between 60 °C and 65 °C, which seems to be the most suitable for the HYBUILD CON concept.

In this temperature range, the commercial product RT64HC manufactured by Rubitherm [19] obtained the best score, followed by the bio-based PureTemp 63 commercial PCM manufactured by the American company Entropy Solution LLC [25]. The third position in the ranking is occupied by another commercial PCM, TH5860, manufactured by German company Terr Hel [32]. The same as for the MED system, shipment costs and other potential additional costs should also be taken into account if a final decision should be made in a specific case, which might affect the ultimate choice for the PCM to be used.

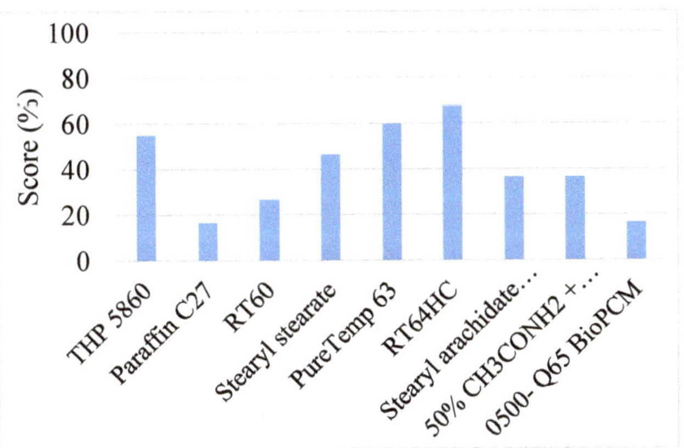

Figure 4. Score obtained by each of the pre-selected PCM candidates in Scenario 1 for the CON system.

The best three PCM candidates, i.e., RT64HC, PureTemp63, and TH5860, were characterized in the laboratory. Figure 5 shows the results of the DSC analysis of different samples after 0, 10, 100, 1000, and 8000 cycles. Both RT64HC and TH5860 show fairly good stability even after 8000 cycles, while PureTemp63 shows an increase in the hysteresis from 3 °C to 6 °C after only 100 cycles. From this point of view, RT64HC and TH5860 are suitable options for this application.

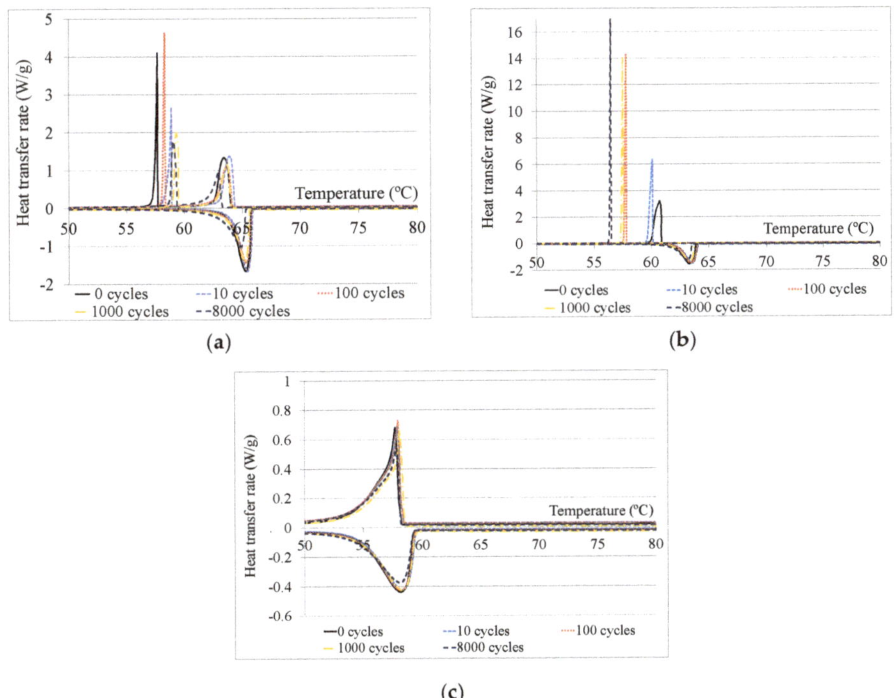

Figure 5. DSC analysis of the cycled PCM. (**a**) RT64HC; (**b**) PureTemp 63; (**c**) THP5860.

The results of the FT–IR analysis performed with the same cycled PCM are shown in Figure 6. No significant changes can be observed in the three PCM as a consequence of the thermal cycling, especially for THP5860, whose spectral curves are practically the same for all cycle numbers.

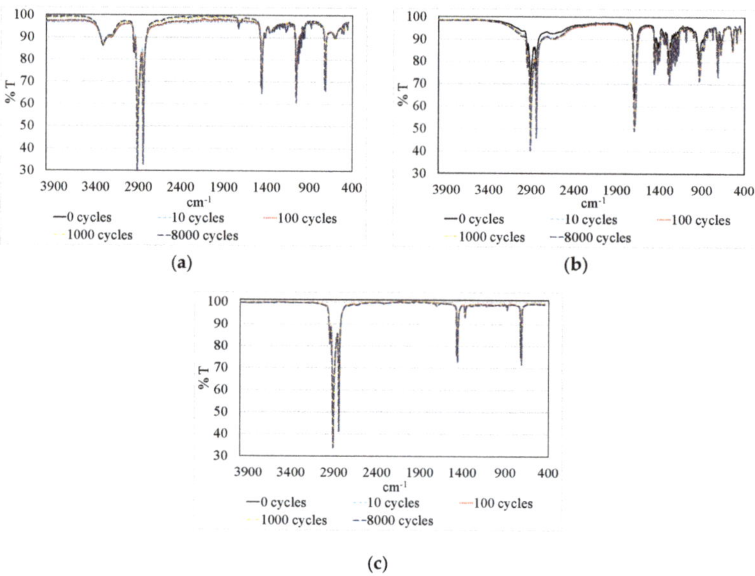

Figure 6. FT–IR analysis of the cycled PCM. (a) RT64HC; (b) PureTemp 63; (c) THP5860.

The results of the TGA analysis of the same PCM candidates as above are shown in Figure 7.

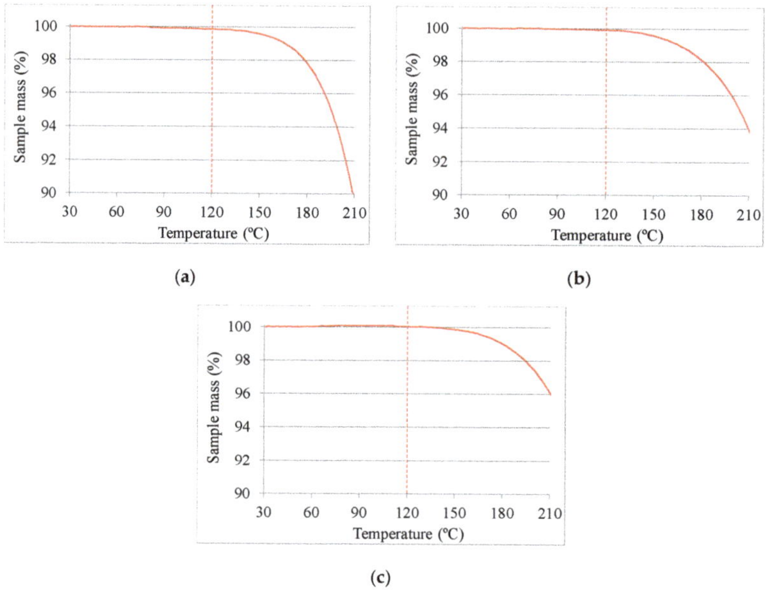

Figure 7. TGA analysis of the cycled PCM. (a) RT64HC; (b) PureTemp 63; (c) THP5860.

The mass loss of RT64HC and PureTemp 63 at 120 °C are 0.13% and 0.09%, respectively, while THP5860 does not lose weight until the temperature reaches around 135 °C. The temperatures at which the mass loss reaches 1.5 wt % are 173 °C, 174 °C, and 188 °C for RT64HC, PureTemp 63, and THP5860, respectively. If a lower heating rate of e.g., 1 °C/min was used instead of 10 °C/min, the temperature for the stability limit would be lower than the values shown here, but still, the mass loss in the temperature range of the application would be negligible.

In summary, according to the decision matrix, the most promising PCM for the CON system, with a phase change temperature between 60 °C and 65 °C, are RT64HC, PureTemp 63, and THP5860. Furthermore, the results obtained by DSC, FT–IR, and TGA confirm that RT64HC and THP5860 are adequate from this point of view, while PureTemp 63 is not indicated because it failed the cycling tests.

3.3. Sensitivity Analysis of the Selection Criteria

The results of the sensitivity analysis are shown in Tables 5 and 6 for the MED and CON systems, respectively, for both scenarios considered as explained in Section 2.2. The average values of the total score are shown in both tables along with the standard deviation calculated over the whole set of altered values of the weights.

Table 5. Average score obtained by the PCM candidates for the MED system.

No.	Scenario 1			Scenario 2		
	Commercial Name/Composition	Average Score (%)	Standard Deviation (%)	Commercial Name/Composition	Average Score (%)	Standard Deviation (%)
1	savE OM 03	61.7	4.6	savE OM 03	66.7	4.6
2	RT4	51.7	5.4	RT4	58.4	5.4
3	A3	43.3	7.2	Caprylic acid + lauric acid (9:1 by mol)	50.2	9.6
4	RT3HC_1	41.7	7.2	RT3HC_1	50.1	7.2
5	PureTemp 4	41.7	7.2	PureTemp 4	50.1	7.2
6	Caprylic acid + lauric acid (9:1 by mol)	40.2	9.6	A3	49.9	7.2
7	PCM-PDR03P	33.3	7.1	PCM-PDR03P	41.6	7.1
8	Tetrahydrofuran clathrate hydrate	29.9	8.4	Tetrahydrofuran clathrate hydrate	24.9	8.4
9	0200- Q2 BioPCM	19.9	5.6	0200- Q2 BioPCM	16.6	5.6
10	0200- Q4 BioPCM	19.9	5.6	0200- Q4 BioPCM	16.6	5.6

Table 6. Average values obtained by the PCM candidates for the CON system.

No.	Scenario 1			Scenario 2		
	Commercial Name/Composition	Average Score (%)	Standard Deviation (%)	Commercial Name/Composition	Average Score (%)	Standard Deviation (%)
1	RT64HC	68.4	7.2	RT64HC	73.3	7.2
2	PureTemp 63	60.1	6.7	PureTemp 63	66.7	6.8
3	THP 5860	55.1	7.2	THP 5860	60.0	7.2
4	Stearyl stearate	46.8	8.3	Stearyl stearate	53.3	8.4
5	Stearyl arachidate (C38H76O2)	36.6	7.8	Stearyl arachidate (C38H76O2)	33.3	7.9
6	50% CH3CONH2 + 50% C17H35 COOH	36.6	7.8	50% CH3CONH2 + 50% C17H35 COOH	33.3	7.9
7	RT60	26.8	6.6	RT60	33.3	6.8
8	Paraffin C27	16.6	5.0	Paraffin C27	13.3	5.0
9	0500- Q65 BioPCM	16.6	5.0	0500- Q65 BioPCM	13.3	5.0

Table 5 shows that, according to the average score obtained for all possible combinations, the first and the second-best options are savE OM 03 and RT4, respectively, in both Scenarios 1 and 2. The third place is occupied by A3 in scenario1 and by the mixture caprylic acid + lauric acid (9:1 by mol) in Scenario 2.

For the CON case, the first place is occupied by RT64HC, followed by PureTemp 63 in the second place, and by THP 5860 in the third place in both Scenarios 1 and 2.

The frequency distributions with respect to the places occupied by the best candidates over the whole set of altered values of the weights are shown in Tables 7 and 8, for the MED and CON systems, respectively. As seen in Table 7, savE OM 03 is always the first option in both scenarios, while RT4 is the second-best option in the vast majority of cases in both scenarios. Therefore, it can be concluded that the sensitivity analysis confirms that the results obtained using the decision matrix of the MED concept are reliable and they are practically not affected by reasonably small variations in the weights of the decision parameters.

Table 7. Ranking of the best three PCM candidates and frequency distribution for the MED system.

Position	Scenario 1		Scenario 2	
	Commercial Name/Composition	Frequency	Commercial Name/Composition	Frequency
1st	savE OM 03	49/49 (100%)	savE OM 03	49/49 (100%)
2nd	RT4	47/49 (95.9%)	RT4	45/49 (91.8%)
3rd	A3	23/49 (46.9%)	Caprylic acid + lauric acid (9:1 by mol)	28/49 (57.1%)

Table 8. Ranking of the best three PCM candidates and frequency distribution for the CON system.

Position	Scenario 1		Scenario 2	
	Commercial Name/Composition	Frequency	Commercial Name/Composition	Frequency
1st	RT64HC	172/180 (95.6%)	RT64HC	172/180 (95.6%)
2nd	PureTemp 63	135/180 (75.0%)	PureTemp 63	150/180 (83.3%)
3rd	THP5860	109/180 (60.6%)	THP5860	111/180 (61.7%)

Table 8 shows that for the CON system RT64HC is the first option in the vast majority of cases (95.6%) in both scenarios, while PureTemp 63 is the second-best option in 75% of the cases in Scenario 1, and in 83.3% in Scenario 2. This means that, except for very few cases that correspond to specific combinations of values for the weights of the decision parameters, which deviate from the reference values considered by the authors, the selection of RT64HC as the best option for the CON concept is quite robust and justified.

4. Conclusions

This paper presented the methodology applied as the initial step of an overall evaluation and materials selection process of the most suitable PCM to be implemented as TES in two innovative compact systems aimed to provide heating, cooling, and domestic hot water (DHW) in residential buildings in the Mediterranean and Continental regions in a more efficient way.

A qualitative decision matrix was used in the pre-screening process to assess the potential of the most promising PCM candidates, by assigning a score to a few relevant properties and calculating a total score based on a weighted average of the scores obtained by the single properties of each PCM. The melting enthalpy and temperature range, availability, cost, and maximum allowed working temperature (for the CON concept), were considered as relevant properties in the decision matrix. A sensitivity analysis was also performed to check the robustness of the selection matrix and the influence that variations in the values assigned to the different weights might have on the results.

For the MED system, which requires a PCM with phase change temperature around 4 °C, given the initial information available, RT4, A3, RT3HC_1, and PureTemp 4 were the most promising options out of a total of ten PCM candidates assessed using the decision matrix. A detailed experimental characterization based on DSC and TGA analysis showed that save OM 03 was unacceptable.

Regarding the CON system, for which the desired phase change temperature should be between 60 °C and 65 °C, given the initial information available, RT64HC, PureTemp 63, and THP5860 were the most promising PCM candidates out of a total of nine PCM assessed using the decision matrix. The detailed experimental characterization based on DSC, thermal cycling, and TGA analysis confirmed that these PCMs were suitable options for the CON system.

It is also important to highlight that the selection process described in this study should be viewed as an initial step in an overall evaluation/materials selection process for the described applications.

Since transportation costs and any other potential additional costs were not taken into consideration in the decision matrix, the results presented here could be different, although they can still serve as the basis for deeper analysis for any specific location and characteristics of the building where either the MED or the CON HYBUILD solution is intended to be implemented. Moreover, a further assessment of the actual thermophysical evaluation of the attained materials, as well as investigating the transient performance of the system and the role of the material selected are crucial steps to ensure the good performance of the entire system.

Author Contributions: Conceptualization, G.Z. and L.F.C.; methodology, G.Z. and A.G.F.; lab testing, A.G.F.; software, G.Z.; formal analysis, G.Z., A.G.F., and L.F.C.; investigation, G.Z., and L.F.C.; writing—original draft preparation, G.Z. and A.G.F.; writing—review and editing, L.F.C.; supervision, L.F.C.; project administration, L.F.C.; funding acquisition, L.F.C. All authors have read and agreed to the published version of the manuscript.

Funding: This project has received funding from the European Union's Horizon 2020 research and innovation programme under grant agreement No 768824 (HYBUILD). This work was partially funded by the Ministerio de Ciencia, Innovación y Universidades de España (RTI2018-093849-B-C31 - MCIU/AEI/FEDER, UE). This work was partially funded by the Ministerio de Ciencia, Innovación y Universidades - Agencia Estatal de Investigación (AEI) (RED2018-102431-T). This work is partially supported by ICREA under the ICREA Academia programme. The authors would like to thank the Catalan Government for the quality accreditation given to their research group (GREiA 2017 SGR 1537).

Acknowledgments: GREiA is certified agent TECNIO in the category of technology developers from the Government of Catalonia. The authors would like to thank TER HELL & Co. GmbH company for providing free samples of their PCM, THP5860, and to PLUSS for providing free samples of their PCM, savE OM 03.

Conflicts of Interest: The authors declare no conflicts of interest. The funders had no role in the design of the study; in the collection, analyses, or interpretation of data; in the writing of the manuscript, or in the decision to publish the results.

References

1. Dincer, I.; Rosen, M.A. *Thermal Energy Storage: Systems and Applications*; John Wiley and Sons: Chichester, West Sussex, UK, 2002; ISBN 978-0471495734.
2. European Commission. *European Commission Communication from the Commission to the European Parliament, the Council, the European Economic and Social Committee and the Committee of the Regions and the European Investment Bank*; European Commission: Brussels, Belgium, 2016.
3. Oró, E.; de Gracia, A.; Castell, A.; Farid, M.M.; Cabeza, L.F. Review on phase change materials (PCMs) for cold thermal energy storage applications. *Appl. Energy* **2012**, *99*, 513–533. [CrossRef]
4. Veerakumar, C.; Sreekumar, A. Phase change material based cold thermal energy storage: Materials, techniques and applications—A review. *Int. J. Refrig.* **2016**, *67*, 271–289. [CrossRef]
5. Cabeza, L.F.; Castell, A.; Barreneche, C.; De Gracia, A.; Fernández, A.I. Materials used as PCM in thermal energy storage in buildings: A review. *Renew. Sustain. Energy Rev.* **2011**, *15*, 1675–1695. [CrossRef]
6. Zhang, Y.; Zhou, G.; Lin, K.; Zhang, Q.; Di, H. Application of latent heat thermal energy storage in buildings: State-of-the-art and outlook. *Build. Environ.* **2007**, *42*, 2197–2209. [CrossRef]
7. Sharif, M.K.A.; Al-Abidi, A.A.; Mat, S.; Sopian, K.; Ruslan, M.H.; Sulaiman, M.Y.; Rosli, M.A.M. Review of the application of phase change material for heating and domestic hot water systems. *Renew. Sustain. Energy Rev.* **2015**, *42*, 557–568. [CrossRef]
8. de Gracia, A.; Oró, E.; Farid, M.M.; Cabeza, L.F. Thermal analysis of including phase change material in a domestic hot water cylinder. *Appl. Therm. Eng.* **2011**, *31*, 3938–3945. [CrossRef]
9. Mazman, M.; Cabeza, L.F.; Mehling, H.; Nogues, M.; Evliya, H.; Paksoy, H.Ö. Utilization of phase change materials in solar domestic hot water systems. *Renew. Energy* **2009**, *34*, 1639–1643. [CrossRef]
10. Xu, B.; Li, P.; Chan, C. Application of phase change materials for thermal energy storage in concentrated solar thermal power plants: A review to recent developments. *Appl. Energy* **2015**, *160*, 286–307. [CrossRef]
11. Prieto, C.; Cabeza, L.F. Thermal energy storage (TES) with phase change materials (PCM) in solar power plants (CSP). Concept and plant performance. *Appl. Energy* **2019**, *254*, 113646. [CrossRef]
12. Zalba, B.; Marín, J.M.; Cabeza, L.F.; Mehling, H. Review on thermal energy storage with phase change: Materials, heat transfer analysis and applications. *Appl. Therm. Eng.* **2003**, *23*, 251–283. [CrossRef]

13. Farid, M.M.; Khudhair, A.M.; Razack, S.A.K.; Al-Hallaj, S. A review on phase change energy storage: Materials and applications. *Energy Convers. Manag.* **2004**, *45*, 1597–1615. [CrossRef]
14. Mehling, H.; Cabeza, L.F. *Heat and Cold Storage with PCM. An up to Date Introduction into Basics and Applications*, 1st ed.; Springer: Berlin/Heidelberg, Germany, 2008; ISBN 978-3-540-68556-2.
15. Miró, L.; Barreneche, C.; Ferrer, G.; Solé, A.; Martorell, I.; Cabeza, L.F. Health hazard, cycling and thermal stability as key parameters when selecting a suitable phase change material (PCM). *Thermochim. Acta* **2016**, *627–629*, 39–47. [CrossRef]
16. Gasia, J.; Martin, M.; Solé, A.; Barreneche, C.; Cabeza, L.F. Phase Change Material Selection for Thermal Processes Working under Partial Load Operating Conditions in the Temperature Range between 120 and 200 °C. *Appl. Sci.* **2017**, *7*, 722. [CrossRef]
17. HYBUILD. Available online: http://www.hybuild.eu/ (accessed on 4 February 2020).
18. Frazzica, A.; Palomba, V.; Sergi, F.; Ferraro, M.; Cabeza, L.F.; Zsembinszki, G.; Oró, E.; Karellas, S.; Varvagiannis, S.; Emhofer, J.; et al. Dynamic Modelling of a Hybrid Solar Thermal/electric Energy Storage System for Application in Residential Buildings. In Proceedings of the ISES EuroSun 2018 Conference—12th International Conference on Solar Energy for Buildings and Industry, Rapperswil, Switzerland, 10–13 September 2018; p. 12.
19. Rubitherm RT-PCM. Available online: https://www.rubitherm.eu/en/index.php/productcategory/organische-pcm-rt (accessed on 2 October 2019).
20. PCM Products. Available online: http://www.pcmproducts.net/ (accessed on 15 January 2019).
21. PCES Phase Change Energy Solutions. Available online: https://phasechange.com/ (accessed on 15 January 2019).
22. RGEES LLC. Available online: http://www.rgees.com/products.php (accessed on 15 January 2019).
23. PLUSS ®. Available online: http://pluss.co.in/ (accessed on 15 January 2019).
24. Shengli, T.; Dong, Z.; Deyan, X. Experimental study of caprylic acid/lauric acid molecular alloys used as low-temperature phase change materials in energy storage. *Energy Conserv.* **2005**, *6*, 45–47.
25. PureTemp LLC. Available online: http://www.puretemp.com/ (accessed on 15 January 2019).
26. Jankowski, N.R.; McCluskey, F.P. A review of phase change materials for vehicle component thermal buffering. *Appl. Energy* **2014**, *113*, 1525–1561. [CrossRef]
27. Pereira da Cunha, J.; Eames, P. Thermal energy storage for low and medium temperature applications using phase change materials—A review. *Appl. Energy* **2016**, *177*, 227–238. [CrossRef]
28. Pielichowska, K.; Pielichowski, K. Phase change materials for thermal energy storage. *Prog. Mater. Sci.* **2014**, *65*, 67–123. [CrossRef]
29. Climator. Available online: http://climatoribiza.com/ (accessed on 15 January 2019).
30. Kenisarin, M.; Mahkamov, K. Salt hydrates as latent heat storage materials: Thermophysical properties and costs. *Sol. Energy Mater. Sol. Cells* **2016**, *145*, 255–286. [CrossRef]
31. Sharma, A.; Tyagi, V.V.; Chen, C.R.; Buddhi, D. Review on thermal energy storage with phase change materials and applications. *Renew. Sustain. Energy Rev.* **2009**, *13*, 318–345. [CrossRef]
32. TER HELL & Co. GmbH. Available online: https://www.terchemicals.com (accessed on 13 January 2020).

MDPI
St. Alban-Anlage 66
4052 Basel
Switzerland
Tel. +41 61 683 77 34
Fax +41 61 302 89 18
www.mdpi.com

Applied Sciences Editorial Office
E-mail: applsci@mdpi.com
www.mdpi.com/journal/applsci

www.ingramcontent.com/pod-product-compliance
Lightning Source LLC
LaVergne TN
LVHW070622170726
843515LV00004B/149

* 9 7 8 3 0 3 6 5 0 8 6 4 1 *